IMMUNOCHEMISTRY OF ENZYMES
AND THEIR ANTIBODIES

DEVELOPMENTS IN
MEDICAL MICROBIOLOGY AND
INFECTIOUS DISEASES

Series Editor: MILTON R. J. SALTON
Chairman, Department of Microbiology
New York University School of Medicine

Alan W. Bernheimer
Mechanisms in Bacterial Toxinology

Milton R. J. Salton
*Immunochemistry of Enzymes
and Their Antibodies*

IMMUNOCHEMISTRY OF ENZYMES AND THEIR ANTIBODIES

Edited by

MILTON R. J. SALTON, Ph.D., Sc.D.
New York University School of Medicine

A Wiley Medical Publication

JOHN WILEY & SONS

New York • London • Sydney • Toronto

Library of Congress Cataloging in Publication Data:

Main entry under title:

Immunochemistry of enzymes and their antibodies.

(Developments in medical microbiology and infectious
diseases) (A Wiley medical publication)
   Includes bibliographical references.
   1. Enzymes—Analysis.   2. Antigen-antibody
reactions.   3. Immunochemistry.   4. Immunospecificity.
I.   Salton, Milton R. J.

QP601.I475   574.1′925   76-28211
ISBN 0-471-74995-8

Printed in the United States of America

10 9 8 7 6 5 4 3 2 1

CONTRIBUTORS

Ruth Arnon, Ph.D., Department of Chemical Immunology, The Weizmann Institute of Science, Rehovot, Israel

Jack London, Ph.D., Laboratory of Microbiology and Immunology, National Institute of Dental Research, National Institutes of Health, Bethesda, Maryland

E. Margoliash, M.D., Department of Biochemistry and Molecular Biology, Northwestern University, Evanston, Illinois

Martin S. Nachbar, M.D., Departments of Medicine and Microbiology, New York University School of Medicine, New York, New York

Joel D. Oppenheim, Ph.D., Department of Microbiology, New York University School of Medicine, New York, New York

Peter Owen, Ph.D., Department of Microbiology, New York University School of Medicine, New York, New York

M. H. Richmond, Ph.D., Sc.D., Department of Bacteriology, University of Bristol, Bristol, England

Stuart Smith, Ph.D., Bruce Lyon Memorial Research Laboratory, Children's Hospital Medical Center of Northern California, Oakland, California

Cyril J. Smyth, Ph.D., Department of Microbiology, New York University School of Medicine, New York, New York

G. J. Urbanski, Ph.D., Department of Biochemistry and Molecular Biology, Northwestern University, Evanston, Illinois

PREFACE

The basis for resistance to infectious diseases has intrigued man for a very long time and especially since the era of the early vaccination experiments of Hunter and Jenner. Immunity to microbial infections remained something of a mystery, even though the use of immunological reactions became very powerful in the definitive diagnosis of diseases and in the determination of host responses to invading microorganisms. At a time when the chemical reactions performed by cells became explicable in terms of specific enzyme protein catalysts, the molecular basis of immunity and immunological reactions remained unexplained. Indeed, during this early period in the development of immunology, immunological phenomena seemed quite bizarre and were enshrouded in an over-complicated terminology. With the introduction of precise chemical, biochemical and analytical approaches to immunology by Landsteiner and Heidelberger and the recognition of specific serum immunoglobulins as distinct physicochemical entities by Tiselius and Kabat, the way was paved for a totally rational and molecular understanding of immunity and the interactions between host-generated antibodies to the invading microorganisms and to the great variety of "foreign" cellular products and toxins that they generate. The discovery by Landsteiner that relatively small chemical entities in a complex molecule could confer specific immunogenicity and behave as determinant groups in an antigen was of tremendous significance. The application of the exact analytical disciplines of chemistry and biochemistry to immunology by Landsteiner, Heidelberger and others made immunochemistry as precise a science as any involved in the study of human response to infectious agents, their toxins and cellular components. The fact that antibodies can be formed to biologically reactive or catalytic protein molecules such as enzymes or cytochromes adds a further exciting dimension to the study of their interactions. Enzymes, unlike many other immunogens lacking intrinsic catalytic reactivity, lend themselves readily to unique combinations of biochemical and immunochemical approaches to studying their interactions. This volume is devoted to selected models illustrating the principles and potentials of such reactions.

As Dr. Herman N. Eisen has aptly pointed out, much of the explosive growth of immunology has derived "from a change in the balance of scientific trade. Until recently, immunology imported more techniques and concepts from related disciplines than it exported. This negative balance is being rapidly redressed: there is now hardly a field of biology or medicine in which antibodies are not useful as analytical or diagnostic reagents, etc." (Herman N. Eisen, *Immunology,* Harper and Row, 1974.) Thus the studies of the interaction of antibodies and immunogenic enzymes has, for example, opened new insights into the molecular basis of their immunological specificities,

vii

taxonomic relationships of enzymes at the molecular level, the molecular topography of cytochrome c, the use of antibodies in localization of membrane enzymes (e.g. ATPase) and the identification and quantitation of enzymes by immunoelectrophoresis.

In this volume we have brought together some selected model systems for studying the interactions of enzymes and their antibodies and for indicating how such investigations can reveal important taxonomic relationships in protein structure (e.g. fatty acid synthetases, malic enzymes and fructose diphosphate aldolase, β-lactamases and cytochrome c), molecular topography of antigens (lysozyme and cytochrome c), the enzymatic architecture of membranes by immunoelectron microscopy (bacterial ATPase) and the tremendous sensitivity and diagnostic potential of enzyme analysis by quantitative immunoelectrophoresis. This latter procedure has not only been invaluable in studying antigenic architecture of cell surfaces and membranes in terms of specific enzymes but has also indicated great diagnostic potentialities for the identification of enzymes by zymogram staining of enzyme immunoprecipitates.

No attempt has been made to make this a comprehensive compendium of all the reported interactions of antibodies and enzymes. Rather, we have selected well-defined and characterized molecular systems to emphasize the principles and potentials of investigating the immunochemistry of enzymes and their antibodies. It is hoped that this volume, devoted to this important area of immunochemistry, will serve to stimulate interest in this field of immunology, which has received relatively little attention since the excellent review of Dr. B. Cinader (B. Cinader, Ed., *Antibodies to Biologically Active Molecules,* Pergamon Press, Oxford) in 1967. Because of the development of the crossed immunoelectrophoresis technique with its great versatility and application to specific identification of enzymes it is clear that important advances will be made through its use in studying enzymes and antibodies to them in human disease. There are many fascinating aspects of enzymes and their antibodies left to be explored. For example the role of human antibodies to important microbial enzymes such as the β-lactamases or other exocellular antibiotic-destroying enzymes has not been investigated. Are such human antibodies of any significance in antibiotic chemotherapy? It is hoped that the selection of enzyme antibody model systems selected for basic characterization will stimulate further investigations of this fascinating area of immunochemistry.

In preparing this volume the enthusiasm and willingness of the contributors has been tremendously encouraging and a great stimulus. I wish to especially thank Mrs. Josephine Markiewicz for all her expert secretarial help in preparing the index and assembling this volume, and Ruth Wreschner and her colleagues at John Wiley and Sons for their encouragement, interest and cooperation. I am also very grateful to those who have contributed material for the authors of the various chapters.

August, 1976   Milton R. J. Salton

CONTENTS

CHAPTER 1

═══════════════════

IMMUNOCHEMISTRY OF LYSOZYME

RUTH ARNON

═══════════════════

1. INTRODUCTION

Significant progress has been made in recent years in immunochemical investigations of macromolecules, particularly in the study of the various parameters of the immune response toward protein antigens at the molecular level. One of the paramount problems, for whose solution enzymes have been used as an important vehicle, is the elucidation of antigenic specificity determinants and their structural parameters. Efforts in this direction had already been made by Landsteiner (1) in his pioneering studies on the specificity of serologic reactions and were continued through numerous recent investigations with synthetic antigens (2) and protein conjugates (3,4). These studies included attempts to identify antigenic determinants in naturally occurring proteins through comparison of the immunologic cross reaction between antigens of related structure, by investigation of the effects of chemical modification on serologic specificity, and by characterization of immunologically reactive fragments of the antigen (5–9).

The studies with *native* proteins were slowed down in the beginning, mainly because their three-dimensional structures were not completed until the early 1960s. Since that time, attempts to delineate the antigenic structure of proteins were carried out in several laboratories and culminated, perhaps, in the recently reported complete mapping of the antigenic structure of myoglobin (10).

The use of *enzymes* rather than other proteins, as antigens, in this type of investigation offers an advantage, because the antigens possess biologic activity that resides in a limited area of the molecule, and antibodies specific toward this or related regions may have an effect on the catalytic activity. Consequently, the contribution of different determinants to the immunologic reactivity of an enzyme can be evaluated in the light of their relationship to its catalytic site.

All enzymes are proteins, namely, multideterminant antigens, and no information is a priori available on the particular groups, or arrangements of groups, that elicit antibody production or serve as points of recognition by the antibodies. Attempts have been made to identify such determinants by comparing the immunologic cross reaction between homologous or isofunctional enzymes, by investigation of the effects of chemical modifications on enzymatic activity and serologic specificity, and by characterization of immunologically reactive fragments that can be prepared from them (reviewed in Ref. 11). It should be borne in mind, however, that the results of every such investigation depend not only on the antigen used but also on the antiserum. The presence of many specificity determinants on the same antigen brings about inevitable heterogeneity in the antibodies, and different individual antisera will, therefore, differ in their potential capacity to react with the various antigenic sites of the molecule. To overcome or bypass this difficulty, it is desirable to have means for differentiation among antisera that vary in the distribution of antibodies with distinct specificities toward different antigenic determinants of the same multideterminant antigen, or for deliberate preparation of antisera specific exclusively to defined regions of the molecule.

Lysozyme is an enzyme particularly suitable for such investigations for several reasons: its three-dimensional structure is recognized in detail (12,13), and its mode of action has also been elucidated (14). Lysozymes from many species have been isolated and analyzed, including the determination of their amino acid sequence. Therefore, clarification of their immunologic relationships may yield information, on a molecular level, concerning their detailed antigenic structure and its changes through evolution (15).

The present article summarizes the information available on the antigenic structure of lysozyme, as derived both from studies of the native enzyme and its derivatives and fragments and from investigation of synthetic analogs that correspond to its fragments. Although for lysozyme it is not possible to delineate in detail all of the antigenic determinants of the molecule, as has been done for myoglobin (16), the accumulated evidence from several approaches does yield a fairly complete picture of the antigenic makeup of the molecule.

2. APPROACHES USED IN IMMUNOCHEMICAL INVESTIGATIONS

2.1 Chemical Modifications

One of the approaches for delineation of antigenic structure is chemical modification of the molecule, either by a treatment that influences its overall conformation or by specific modification of particular amino acid side chains. The latter may yield information on the extent to which particular residues in the molecule contribute to the antigenic specificity, but in some cases, the chemical modification will result also in a change in conformation of the enzyme, an effect that will contribute to the change in antigenic reactivity.

In the case of lysozyme, several modified derivatives were tested for their antigenic properties, mainly in correlation with the enzymatic properties. Special attention was paid to the tryptophan residues, several of which are known to be directly involved in the catalytic site. Modification of all six tryptophan residues with 2-nitrophenylsulfenyl chloride destroys the enzymatic activity completely, concurrently with a drastic decrease, by 82%, of its ability to react with antibodies to the native enzyme (17). The large reduction in the immunologic activity of this derivative might reflect actual involvement of one or more of the tryptophan residues in antigenic reactive regions or might simply be the result of the conformational changes that were shown to occur on modification. Modification of five tryptophan residues, by formylation, caused a less drastic change in antigenic reactivity (18). When only one tryptophan residue (no. 62) of lysozyme was modified by reaction with N-bromosuccinimide, the derivative, although enzymatically inactive, was fully reactive with antilysozyme antibodies, as was another derivative, which retained 60% of the catalytic activity, in which either tryptophan residue 62 or 63 was modified (19). These findings indicate that the tryptophan residue side chains per se do not contribute to a large extent to the antigenic reactivity of lysozyme.

On the other hand, the tyrosine residues participate in the immunologic activity: modification of three of the six tyrosine residues reduced the immunologic capacity, although it did not impair the enzymatic activity (19). Modification by nitration of only two tyrosine residues (nos. 20 and 23), which was accompanied by a conformational change, also brought about a slight decrease in antigenic activity (2).

As for the ε-amino groups of the lysine residues, guanidization, maleylation, or succinylation of all of the amino groups (21,22), or their modification by acetylation or carbamylation (23), resulted in a considerable decrease in the ability to react with the antibodies to native lysozyme and, in parallel, brought about an even more drastic decrease in the catalytic activity. Habeeb and Atassi (24) reported that on reversible blocking of all of the amino groups in lysozyme by reagents like tetrafluorosuccinic anhydride or citraconic anhydride, the derivatives lost all of the enzymatic activity and about 60% of their antigenic activity. Both activities could be restored on deblocking and release of the free amino groups.

Modification of carboxyl residues was attempted by coupling with two reagents, glycine methyl ester and histidine methyl ester. In both cases, only two carboxyls were modified, namely, that of aspartic 119 and the carboxy terminus (25). The attachment of histidine methyl ester, which is accompanied by conformational changes (26), caused a marked decrease in the antigenic activity, whereas the glycyl derivative was antigenically identical to native lysozyme. These results indicate that aspartic 119 and leucine 129 are not part of an antigenically reactive region.

Another amino acid side chain that was tested for its participation in antigenic activity is arginine. Modification of all 10 arginine residues by reaction with cyclohexanedione resulted in the loss of all enzymatic activity and almost all antigenic activity. This loss was probably due to drastic conformational changes in the molecule. When only one arginine residue (no. 61) was modified with phenylglyoxal, the resultant derivative, with a conformation almost identical to that of intact lysozyme, retained most of its enzymatic activity and was antigenically identical to lysozyme (27). These findings indicate that Arg 61 does not form part of an antigenic determinant, but they do not provide any information about the contribution of the other arginine residues, because the above-mentioned loss of antigenic reactivity could have resulted from the change of conformation alone.

From the accumulated information above, which is summarized in Table 1, it may be concluded that chemical modification of particular amino acid side chains has only limited value in pinpointing the antigenically reactive regions. This is due mainly to the fact that usually more than one residue of a kind is modified during the reaction and also that, except in very few cases, when only a single residue is modified, the effect on antigenicity is rather small.

2.2 Fragmentation

An alternative approach for the identification of antigenically reactive regions of a protein is fragmentation of the protein either by chemical cleavage adjacent to specific residues or by limited proteolysis. This procedure is followed by fractionation of the resultant fragments and their screening for immunologically active components, namely, components that can bind to the antibodies or interfere with the interaction of the antibodies and the intact antigen. Because such fragments, by definition, embody antigenic

Table 1. Effect of Chemical Modification on Catalytic and Antigenic Activity of Lysozyme

Modified residues	Catalytic activity (%)	Antigenic reactivity[a] (%)	Reference
Tryptophans (all 6)	0	18	17
Tryptophans (5 of 6)	0	85	18
Trp_{62} or Try_{63}	60	Full	19
Trp_{62}	0	Full	19
Tyrosines (3 of 6)	Full	Limited decrease	19
Tyr_{20} and Tyr_{23}	50	77–90	20
Lysines	0	60	21–24
Asp_{119} and Leu_{129}	27	Full	25
Arginines (all 10)	0	8–22	27
Arg_{61}	83	Full	27

[a] Extent of cross reactivity with antilysozyme antibodies.

determinants, and can be subsequently analyzed and defined, they have been utilized to characterize the antigenic structure of the parent molecule. Furthermore, by subjecting them to various chemical modifications, one can identify the particular residues that contribute to the antigenic reactivity and specificity. This technique, although very useful, is not without its shortcomings. The most serious objections are the following:

1.   An isolated peptide will not necessarily maintain the conformation it had in the parent protein (28,29). For this reason, a nonreactive fragment may, indeed, still be a part of a reactive region in the intact protein. However, with some reactive region-carrying fragments, the antibody to the intact molecule may induce on the peptide the conformation necessary for binding with the antibody-combining site (30).

2.   Cleavage of the protein may occur *within* a reactive region. The consequence of dissecting the reactive region into two (or more) different peptides may be the loss of most of the immunochemical reactivity of the site. This is a serious objection, because it may be difficult to locate *all* of the antigenic determinants, unless several cleavage techniques are used for the same protein and overlapping fragments are analyzed.

Notwithstanding these shortcomings, however, this technique is still one of the better approaches available for the study of antigenic determinants. When positive results are obtained with a given peptide, the data are useful, particularly if the same region is isolated as a part of several different overlapping peptides. As will be clear from Sec. 4, most of our knowledge about the antigenic makeup of lysozyme is derived from the investigation of immunologically reactive regions obtained by fragmentation.

2.3   Sequential versus Conformational Determinants

The antigenic determinants of proteins have been divided on a theoretical basis into two broad categories, "sequential" and "conformational," depending on whether their specificities are a function of the amino acid sequence only or of the conformation of the whole protein (31). According to this classification, a sequential determinant was defined as one due to a segment in the amino acid sequence in its unfolded, or linear, conformation, and antibodies to such a determinant are expected to react with a peptide of identical or similar sequence. On the other hand, a conformational determinant was defined as one that results from the steric conformation of the antigenic macromolecule and leads to antibodies that would not necessarily react with peptides derived from that area of the molecule. Thus, conformational determinants would include those determinants composed of amino acid residues that are remote in the unfolded peptide chain but occupy juxtapositions in the native folded structure.

Examination of the three-dimensional structures of several globular proteins reveals that they contain short sequences of adjacent amino acids whose side chains are partially or fully exposed on the surface of the protein, for example, residues 77–80 and 81–85 in sperm whale myoglobin (32) and residues 53–63 of lamprey hemoglobin (33). Consequently, the existence of sequential determinants in globular proteins is at least theoretically feasible. However, in practice, little convincing evidence in support of their occurrence has been reported. Indeed, the almost complete loss of antigenicity on denaturation argues strongly in favor of the view that the vast majority of the determinants that elicit humoral antibody formation are conformational.

3.   ROLE OF CONFORMATION IN ANTIGENICITY OF LYSOZYME

As mentioned above, conformation plays a crucial role in the antigenic specificity of most globular proteins. The three-dimensional shape of protein molecules is dictated by the amino acid sequence (primary structure) and the unique folding of the linearly synthesized polypeptide chain. The conformation is established by noncovalent interactions between the amino acid residues of the polypeptide chain, such as hydrogen bonding, which gives rise to α-helix and β-structure (secondary structure), and is stabilized by intramolecular covalent cross-links, that is, disulfide bridges (tertiary structure). Disruption of the secondary and/or tertiary structure usually results in loss of antigenicity.

The immunologic specificity of lysozyme is almost entirely dependent on its three-dimensional native conformation. This dependence has been implied mainly from the finding that cleavage of the disulfide bonds by reduction and carboxymethylation, and consequent unfolding of the polypeptide chain, caused the enzyme to lose its capacity to react with the antibodies. The reduced and carboxymethylated lysozyme was by itself immunogenic; however, it did not cross react with antibodies to native lysozyme, nor did lysozyme cross react with the antibodies it elicited (34,35). This is true as far as the humoral antibody response is concerned. On the cellular level, these two molecules exhibit a certain degree of cross reactivity (36), but this effect is due more to the differences between cellular recognition and humoral response than to the antigenic similarity between the native enzyme and its unfolded polypeptide chain; this phenomenon will be discussed in detail in a later section.

In addition to the disruption of the disulfide bonds, the process of reduction and carboxymethylation also elicits like-like charge repulsion of the carboxymethyl anions, which prevents the regions previously linked by the disulfide bond from reapproaching each other. To determine which of the two factors is decisive for the antigenic reactivity, this reduced and carboxymethylated lysozyme was compared with another derivative, in the preparation of which the reduction was followed by methylation of the sulfhydryl groups with methyl-*p*-nitrobenzenesulfonate (37). The results indicated that although both derivatives were greatly unfolded, the methylated derivative retained somewhat more of its folded conformation, as indicated by optical rotatory dispersion and circular dichroism measurements. Concomitantly, it retained 25–40% of the capacity to react with antilysozyme serum, whereas the reduced and carboxymethylated derivative showed no reaction (0%) with this antiserum. It must be concluded, therefore, that after disruption of the disulfide bonds, the molecule may partially refold, because the folding is a direct consequence of the primary sequence (38–43) and the longrange intermolecular interactions (44). However, in the carboxymethylated derivative, the mutual repulsion of the relevant regions may hinder the refolding and the consequent restoration of the antigenic capacity.

4. DEFINED ANTIGENIC DETERMINANTS OF LYSOZYME

Upon enzymatic cleavage, it is feasible to obtain from lysozyme fragments that possess immunologic activity, that is, contain antigenic determinants. The susceptibility of lysozyme to tryptic digestion is extremely low, as is that of most native disulfide bond-containing "tight" proteins. However, its digestion by pepsin yields several peptides of different sizes (45,46). Separation of these fragments and their screening for interaction with antilysozyme antibodies has made possible the localization of several antigenic determinants in lysozyme.

4.1 The Region of Residues 57–107

In the first studies on isolation of antigenic fragments of lysozyme (45), it was observed that on digestion of lysozyme with pepsin at pH 1.62, two polypeptides could be isolated, which were capable of inhibiting the precipitin reaction of lysozyme with its specific antiserum but were incapable of inhibiting the neutralization of lysozyme by the antibodies. The conclusion was drawn that the antibodies reactive with these polypeptides are nonneutralizing antibodies. In later studies (47), the same authors elucidated the structure of four immunologically active peptides, purified from a peptic digest of lysozyme. They were all located between Gln_{57} and Ala_{107}, but each had a cleavage in the loop between Cys_{80} and Cys_{94}. These four peptides seemed to bear at least one antigenic determinant, and they inhibited the homologous precipitin reaction to a maximum of about 25%, but did not inhibit the neutralization of lysozyme by the antibodies. Another peptide, which was obtained by tryptic degradation and consisted of three peptide sequences, Trp_{62}–Arg_{68}, Asn_{74}–Leu_{83}, and Leu_{84}–Lys_{96} linked by two disulfide bonds, Cys_{64}–Cys_{80} and Cys_{76}–Cys_{94}, was also found to bear weak but definite immunologic activity.

Application of somewhat different conditions for the peptic digestion (46) yielded another fragment of lysozyme that originated from the same region of the molecule and consisted of the sequences Ser_{60} to Leu_{83} and Ser_{91} to Trp_{108} linked by two disulfide bonds, an intrachain Cys_{64}–Cys_{80} and an interchain Cys_{76}–Cys_{94} (48). This peptide was immunologically active: it was capable of inhibiting the homologous precipitin reaction of lysozyme to 25%. Moreover, an immunoadsorbent prepared from this peptide adsorbed between 20 and 30% of the antilysozyme antibodies, a fraction that, though capable of binding radioactively labeled lysozyme, had only low inhibitory effect on the catalytic activity of the enzyme (48).

It can, therefore, be concluded that the region confined by the sequence Gly_{57} to Ala_{107} contains at least one antigenic determinant.

4.2   N-C Terminal Region

In addition to the large fragment just described, peptic digestion yields another relatively large fragment, which also possesses immunologic activity. This fragment consists of the two terminal peptides, that is, Lys_1 to Asn_{27} and Ala_{122} to Leu_{129}, linked together by a single disulfide bond (Cys_6–Cys_{127}). This peptide was isolated by several laboratories (49,50) and was shown to encompass an antigenic determinant that is independent of the one described above. This peptide did not give any precipitin reaction with antilysozyme antibodies, but, as demonstrated by equilibrium dialysis studies, it is capable of binding to these antibodies with an association constant of 1.75×10^5 (49). The fraction of antilysozyme antibody directed against this peptide was evaluated at 47%. The independence of this determinant from the fragment between residues 57 and 107 was also demonstrated by equilibrium dialysis experiments (51), where each peptide was labeled by acetylation with [^{14}C]acetic anhydride, and the effect of the other peptide (unlabeled) on its binding to the antibodies was examined. The results demonstrated that the binding of neither peptide was influenced by the other. This independence of these two immunologically active fragments is in accord with information on the three-dimensional structure of lysozyme, which indicates that the two portions of the molecule that correspond to them are located far apart (14).

4.3   Antigenic Regions Relevant to the Antibody Inhibitory Effect

Antibodies to lysozyme are capable of inhibiting the enzyme activity to a great extent (52). Nevertheless, early studies already indicated that not all antilysozyme antibodies are inhibitory; some antibody fractions are completely noninhibitory and can form, with lysozyme, antigen-antibody complexes that are enzymatically fully active (45). Moreover, it was even demonstrated that selectively separated antibody fractions, which had different inactivating capacities, show less inhibitory efficacy than does their mixture (48). This finding suggests that the inhibitory efficiency of the original unfractionated antibodies is due to simultaneous combination of lysozyme with several antibody molecules of differing specificities.

However, there is evidence that fractionation of antibodies may also lead to a highly inhibitory fraction that was isolated by means of an interesting approach. Based on the findings (53) that a small-molecular-weight lysozyme inhibitor, tri-N-acetylglucosamine, partially inhibited the serologic activity of lysozyme, the same inhibitor was used for the dissociation of lysozyme-antilysozyme complexes (54). The antibodies fractionated in this manner (which comprised 7–8% of the total precipitating antibody) were efficient inhibitors, even with regard to the activity of lysozyme on low-molecular-weight substrates. Although this antibody fraction was still precipitable with lysozyme, an effect that implies that these antibodies are not specific to the substrate binding site exclusively, a selective fractionation based on inhibitory capacity was indeed achieved in this case, resulting in antibodies toward a limited number of antigenic sites.

In more recent studies from the same laboratory (55), a peptide was isolated, denoted P_{Ib}, antibodies to which had high efficiency in neutralization of the enzymatic activity of lysozyme for both the large substrate *Micrococcus lysodeikticus* and the small substrate hexa-*N*-acetylchitohexaitol, comparable with the inhibition effected by unfractionated antilysozyme antibodies. This immunologically active peptide was isolated after limited digestion of lysozyme by thermolysin. It consists of two peptides, namely, sequences 29–54 and 109–123, linked by a disulfide bond (Cys_{30}–Cys_{115}). The antibodies specific for this peptide were isolated from antilysozyme by the use of the peptide-Sepharose as an immunoadsorbent and were shown to have an association constant to 1.84×10^5 with lysozyme. Inhibition studies indicate that the specificity of these antibodies is directed mainly at the region 38–54, which includes the residue Glu_{52}, known to be part of the lysozyme catalytic site, thus rationalizing their high inhibitory capacity.

4.4   Fragments Obtained by Tryptic Digestion of Lysozyme

Native lysozyme is a "tight" globular protein; that is, it is inaccessible to proteolytic attack with its disulfide bonds intact. The molecule has, therefore, to be appropriately treated to render it accessible to tryptic hydrolysis. One possible treatment was the reversible blocking of the amino groups of lysine with citraconic anhydride (56). The protein was then hydrolyzed by trypsin at all arginine residues. After deblocking, it was possible to effect complete cleavage of all of the lysine peptide bonds, thus yielding all of the tryptic peptides of lysozyme, without rupturing the disulfide bonds. This mixture of peptides inhibited the reaction of lysozyme with its antibodies very strongly (85–89%). The reactive fragments were identified as consisting chiefly of the three disulfide-containing peptides, which comprise the following sequences: 22–33 and 115,116 linked by Cys_{30}–Cys_{115}, 62–68 and 74–96 linked by Cys_{64}–Cys_{76}–Cys_{94}, and 6–13 and 126–129 linked by Cys_6–Cys_{127}. No report from that laboratory is available yet on the contribution of each of these peptides to the total antigenic reactivity in lysozyme.

Fujio and collaborators (57) have isolated one of these fragments, which consists of sequences 62–68 and 74–96, linked by two disulfide bonds, and denoted it P loop II. About 5–10% of antilysozyme antibodies reacted with this peptide, which had distinct specificity from loop 64–80. These antibodies had an association constant of 1.1×10^5 with lysozyme, and they could neutralize its enzymatic activity when *M. lysodeikticus* was used as the substrate, but not for the small substrate hexa-*N*-acetylchitohexaitol. These antibodies are therefore not involved with the active site itself but exert their inactivation by steric hindrance.

4.5   The Loop Region of Lysozyme (residues 64–80)

Peptides from region 57–107, which were mentioned earlier (Sec. 4.1) can yield a smaller peptide that still retains immunologic activity (58). This fragment, which consists of the amino acid sequence 60–83 and contains one intrachain disulfide bond (Cys_{64}–Cys_{80}), was designated as the "loop." The location of this region in the three-dimensional structure of lysozyme is shown in Fig. 1. Antibodies specific to this region only have been prepared by two alternate procedures. One of these techniques utilized a specific immunoadsorbent, which contained the loop peptide attached to a solid support, for selective isolation of antibodies from antilysozyme serum. The alternate procedure consisted of immunization of rabbits and goats with a conjugate that contained the loop

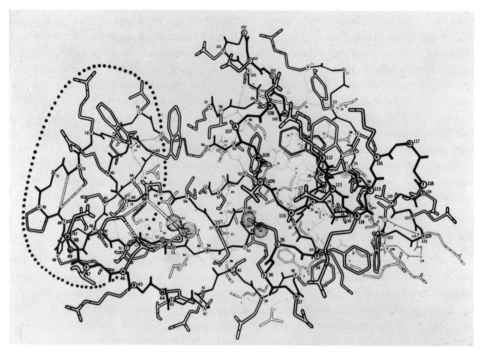

Fig. 1. Backbone and side chains of lysozyme (from the Atlas of Protein Sequence and Structure. M. O. Dayhoff, Ed.). The loop region is encircled.

peptide bound to a synthetic carrier (multi-poly-DL-alanyl-poly-L-lysine, abbreviated A—L), that elicited the formation of antibodies reactive with lysozyme. By adsorption on, and elution from, a lysozyme immunoadsorbent, a preparation of antibodies was obtained, with a specificity directed against a unique region in native lysozyme. The antiloop antibodies obtained by either of these two procedures showed, as expected, less structural heterogeneity than did the totality of the antilysozyme antibody population. This finding was manifested both by acrylamide electrophoresis of the respective light chains and by isoelectric focusing of the intact antibodies (59).

The detailed specificity of antiloop antibodies was investigated by three different sensitive techniques: antigen-binding capacity with radioactively labeled antigens, inactivation of modified bacteriophage (60,61) preparations coated chemically with either lysozyme or the loop peptide, and a fluorometric method that employed a loop derivative to which a fluorescent chromophore was attached; the following results were obtained.

The capacity of the antibodies to bind radioactively labeled antigen was the same for lysozyme and for the isolated loop fragment: in high antibody concentration, both antigens were bound to an extent of 90%. On the other hand, the open-chain loop peptide, in which the disulfide bond had been disrupted by performic acid oxidation, showed only very limited binding (59). The experiments with the modified bacteriophage yielded similar results: bacteriophage preparations, conjugated with either lysozyme or the loop peptide, were efficiently inactivated by the antiloop antibodies. This inactivation was, in turn, inhibited by lysozyme and by the loop peptide. The open-chain peptide, obtained by reduction and carboxymethylation of the loop, was, however, a much weaker inhibitor (59).

The fluorometric method was developed in our laboratory specifically for use on the loop system. For this purpose, the loop peptide was labeled with the dansyl (1-dimethylaminonaphthalene-5-sulfonyl) group as an external fluorophore (62). The resultant fluorescent derivative served as an environmental probe for the interaction of the loop with the antibody. Indeed, binding of antiloop antibodies to this peptide derivative led to a specific excitation energy transfer from the antibody to the dansyl group and was manifested as an apparent fluorescent enhancement, which increased as a function of the amount of antibodies. This enhancement was, in turn, competitively inhibited by the addition of unlabeled loop peptide or by equimolar amounts of intact lysozyme (Fig. 2). The loop peptide was a very efficient inhibitor: it brought about 50% inhibition when added in a ratio of 2:1 to the dansylated loop. The open-chain loop peptide, obtained by reduction and carboxymethylation, did not have any effect on the fluorescence of the dansyl-loop antibody complex.

The data accumulated with the aid of all three techniques are evidence of one and the same phenomenon: the immunologic reactivity of the loop is drastically reduced as a result of unfolding of the peptide chain, a finding that indicates the decisive role played by spatial conformation in the antigenic specificity of this unique region in the lysozyme molecule. Further corroborating evidence for this conclusion will be presented in the next section, which describes synthetic analogs of the loop peptide.

It was mentioned earlier that electrophoresis and isoelectric focusing experiments indicated that the antiloop antibodies showed restricted structural heterogeneity (59). By use of physicochemical methods for measurement of the binding parameters, their functional homogeneity was also demonstrated. Two methods were used for this purpose: equilibrium dialysis with ^{14}C-labeled loop peptide (by attaching a radioactive

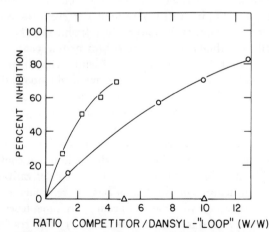

Fig. 2. Inhibition of enhanced fluorescence of the mixture of dansyl-loop (4×10^{-6} M) and antiloop antibodies (0.6×10^{-6} M) by varying concentrations of the loop peptide ($\square$), hen egg-white lysozyme ($\bigcirc$), and the open-chain peptide obtained by reduction and carboxymethylation of the loop peptide ($\triangle$).

carboxymethyl group to the free sulfhydryl group, cysteine residue 76), which pointed to an association constant of 3.0×10^6 M^{-1} and a homogeneity index of 1.02 ± 0.05, and fluorometric measurements, which utilized the interaction between the antiloop antibodies and the dansylated loop derivative described above. By use of the data obtained from the "titration" curve of the increase in fluorescence as a function of antibody concentration, the association constant and the homogeneity index of the antibodies were calculated to be 3.2×10^6 M^{-1} and 0.98 ± 0.05, respectively, in very close agreement with the results obtained in the equilibrium dialysis experiments (62). Both showed a homogeneity index close to unity which is indicative of a high degree of functional homogeneity of these antibodies, which are specific toward a unique conformation-dependent antigenic region of lysozyme.

5. SYNTHETIC LOOP AND ITS ANALOGS: ANTIGENIC SPECIFICITY

Because the loop region between disulfides Cys_{64} and Cys_{80} is a relatively short peptide segment that consists of only 17 amino acid residues with a known sequence, it has been feasible to synthesize it, by chemical methods, with the anticipation that the synthetic approach may yield more detailed information regarding the antigenic specificity or, rather, the molecular requirements for antigenic specificity. The synthesis was carried out by use of the solid-phase technique (63), which yields a loop-like peptide in which the only difference from the natural loop of lysozyme is the replacement of cysteine in position 76 with alanine, to avoid ambiguous disulfide bond formation. The synthesized loop, which corresponds to positions 64–82 in the amino acid sequence of lysozyme, was identical in its properties and immunologic reactivity to the natural loop peptide. It was attached to the same synthetic carrier mentioned above, namely, A—L, and the resultant completely synthetic macromolecule was used for immunization of both rabbits and goats (64). The antibodies it elicited were similar in every respect to those obtained on immunization with conjugates of the natural loop fragment. They were reactive with lysozyme, and their specificity was directed toward a conformation-dependent determinant. Thus, the antibodies reacted very efficiently with either intact lysozyme or the closed loop fragment but did not react with the unfolded loop peptide.

One of the advantages offered by a synthetic approach in biochemical and biologic studies is that after the biologic properties of one synthetic material have been unequivocally demonstrated, many analogs of it may be prepared and tested. Because the chemistry of these analogs is known and controlled, they can lead to an understanding of the role played by different parameters of the molecule in conferring the specific biologic activity. With this point in mind, several analog derivatives of the lysozyme loop were prepared in which one or two amino acids were replaced by alanine, for the purpose of testing the extent of involvement of proline, arginine, leucine, or isoleucine in the immunologic reactivity of this antigenic region (65). The results (Fig. 3), which show the capacity of the various analogs to bind to antiloop antibodies, demonstrate that the derivatives in which either leucine (residue 75) or isoleucine (residue 78) were replaced by alanine were almost indistinguishable from the intact synthetic loop. On the other hand, replacing both proline residues (70 and 79) or only one proline (residue 70) brought about a drastic decrease in the antigenic reactivity. Another analog, in which the two arginine residues (68 and 73) were replaced, also completely lacked the ability to bind to the antibodies, an effect that was attributed to the bulkiness of these side chains (66). It is of interest that when the linear peptides of these loop analogs were tested for their binding to antibodies specific for unfolded reduced carboxymethylated lysozyme, they all showed a similar reduced binding, as compared to the linear peptide of the intact loop, which indicates that in this system, spatial conformation does not play a very important role (66).

More subtle changes in the gross conformation of the molecule also affected its antigenic reactivity. Thus, it was recently observed (67) that reduction of the loop, followed by conversion of the cysteinyl residues to dehydroalanine, thereby forming a linear structure devoid of negative charge at its termini, resulted in a marked decrease in the binding efficiency (Table 2); however, this change was not as drastic as that effected by reduction and carboxymethylation. Reduction followed by reoxidation led to a complete reversal of antigenic activity (59). If, however, the reoxidation was performed in the presence of mercuric chloride, the product of the closure of the loop, which contained one mercury atom per molecule, was slightly less active than the intact loop (Table 2). These data indicate that even a relatively small increase in the size of the loop may change the immunologic properties.

These cumulative findings, therefore, provide evidence of the crucial effect of conformation in the antigenic specificity of native proteins. Whereas in the unfolded polypeptide chain each individual amino acid partakes in the immunologic activity to a similar extent, in the system of the conformation-dependent antigenic determinant, the

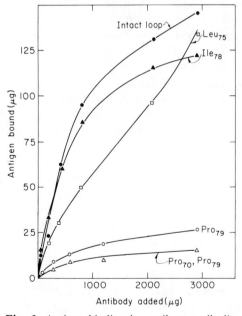

Fig. 3. Antigen binding by antiloop antibodies of the various loop analogs. The designation of the derivatives is according to the amino acid(s) replaced by alanine.

Table 2. Binding Properties of Loop Derivatives to Antiloop Antibodies

Derivative	Observed binding activity (%)
Intact synthetic loop (reduced and reoxidized)	100
RCM loop (reduced carboxymethylated)	0
Linear loop peptide (reduced and cysteines converted to dehydroalanine)	56
Hg loop (reduced and reoxidized in the presence of $HgCl_2$; containing 1 Hg atom per peptide molecule)	88

loop, only residues that have a decisive role in determining the overall shape of the molecule affect its antigenic properties and reactivity.

6. IMMUNOGENICITY OF THE LOOP AND ITS INTERACTION WITH ANTIBODIES

The data presented until now refer not to the immunogenicity but only to the antigenic specificity of the different regions of lysozyme as deduced from the capacity of various molecular fragments to interact with antibodies evoked either by the intact lysozyme molecule or by a conjugate that contains the fragment, the loop, attached to a macromolecular carrier. Attempts to immunize with the isolated loop did not lead to detectable amounts of antibodies either in rabbits or in goats. This phenomenon was attributed to the small size of the loop, because it has been reported that other peptides of a similar size were also immunogenic only when bound to macromolecular carriers (68,69). This finding is also in accord with the currently accepted concept of the nature of T- and B-cell cooperation, which predicts that a theoretical minimum requirement for elicitation of humoral antibody production is the presence of two different antigenic determinants that fulfill the carrier and hapten functions. This requirement may not even be sufficient unless the different determinants are spatially disposed in a way that permits cell-cell interactions (70). It is thus implied that there is a lower limit in the size of the antigen for immunogenicity, which is probably above the size of the loop.

All the same, it was recently observed that immunization of mice of the DBA/1 strain with the loop alone resulted in an immune response, as manifested by the presence of antibody-forming cells in their spleens, at both primary and secondary responses (71). According to this criterion, the response to the loop was comparable, if not higher, than that to lysozyme; however, the titer of circulating antibodies, as determined by passive hemagglutination, was lower in response to the loop than to lysozyme. The involvement of the T-cell population in these responses was indicated by the capacity of the loop to exert a "carrier" effect. It is thus established that the loop as such can be immunogenic.

In the interaction with its antibodies, the loop, likewise, exhibits characteristics, or behavior, different from those of other small-molecular-weight haptens. One aspect in the study of the interaction between antigen and antibody is the possible changes it might induce in conformation of the antibody. Recent investigations, indeed, report on the existence of such changes, as detected by sophisticated physicochemical techniques. Thus, by use of the technique of small-angle x-ray diffraction, it has been possible to demonstrate that intact antibodies to poly-D-alanine do undergo small changes in their volume as a result of interaction with their hapten, tetra-D-alanine (72), a change that is not observed when the hapten reacts with either Fab or (Fab′)₂ fragments of the antibodies (73). Other studies, which employed the technique of circularly polarized luminescence (CPL), detected changes both in intact antibodies and in their Fab fragments as a result of their interaction with their respective antigens (74). By use of high-molecular-weight antigens, such as RNase and poly-DL-alanine, the spectral data demonstrated that the changes in conformation of the Fab fragments were markedly different from those observed with the whole antibody molecules, a finding that indicates interaction between the Fc and Fab fragments of the antibody and, probably, a change in the conformation of Fc. With monovalent haptens, differences were observed between the small phosphorylcholine, which did not cause any change in CPL of the antibody, tetra-D-alanine, the binding of which revealed changes only in the Fab fragment, and the lysozyme loop, which induced alterations in both Fab and the intact antiloop antibody. This finding implies that the loop differs from small haptens in the conformational changes in the antibody that it brings about.

On the other hand, the lysozyme loop, like other monovalent haptens, is not capable of effecting a complement fixation reaction (75), whereas its dimer, prepared by coupling two loop molecules via nonamethylene diamine (H_2N—$(CH_2)_9$—NH_2), fixes complement readily, just like the polyvalent conjugate mentioned above, namely, loop-A—L (Fig. 4). The concomitant influence of the loop dimer on CPL spectra of antiloop antibodies is both different and more extensive than that effected by the loop itself (75), a phenomenon that might be attributed to the capacity of the dimer to form oligomeric complexes with the antibodies (76,77) or to enable the closure of a monomeric ring form (77a) in which an analogous conformational change is involved.

7. RELATIONSHIPS OF LYSOZYMES FROM VARIOUS SPECIES

One of the interesting topics in biochemistry, and mainly in enzyme studies, is the question of biochemical evolution. Immunologic approaches seem to be particularly suitable

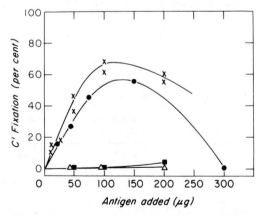

Fig. 4. Complement fixation by antiloop antibodies with loop-A—L conjugate (×), loop-dimer (●), loop (■), and lysozyme (△).

for such studies, because antibodies to specific enzymes can be employed in the search for enzymes of biologic pathways that disappeared in the course of evolution or to detect the extent of similarity between enzymes that persisted through the ages. In cases where the primary structure of the enzyme is known, the elucidation of the immunologic behavior is feasible in precise molecular terms and should provide a sensitive probe for studying the surface conformation. Thus, comparative immunochemical studies of lysozymes of various species have contributed to our understanding of both evolutionary problems and of structural concepts relevant to this enzyme.

7.1   Evolutionary Aspects

In general, the process of evolution has been in the direction of greater diversity of the enzymes both in chemical structure and in biologic specificity. Whereas the latter leads to the occurrence of homologous enzymes with different specificities, such as trypsin and chymotrypsin, the former leads to the existence of equifunctional enzymes in many species or in various organs of the same species. The relationships between such phylogenetically homologous enzymes involve not only chemical similarity but also conformational homology, for which immunologic cross reaction may be taken as corroborating evidence.

For the assessment of the relatedness of lysozymes from different species, it is most valuable to determine their amino acid sequences and three-dimensional structures. However, it is not always desirable to resort to such tedious and expensive methods. Certain immunologic methods can far more easily provide estimates of the approximate degree of sequence resemblance and conformational similarity among related lysozymes.

Indeed, lysozymes from 16 species of birds were compared for their degree of interaction with antiserum to hen egg-white lysozyme and with antisera prepared against lysozymes purified from egg whites of turkey, Japanese quail, bobwhite quail, and duck. With the complement fixation method, all of the bird lysozymes tested were found to be related to each other (78–80), as demonstrated in Table 3.

With antihen egg-white lysozyme, the strongest cross reaction was obtained with quail lysozyme, while pheasants were at the weaker end of the reactivity series. These results were unexpected because according to zoologic evidence, pheasant is closer to hen; however, a subsequent comparison of the amino acid sequences of these lysozymes revealed only four amino acid interchanges between hen and quail lysozymes, compared to seven interchanges in the pheasant enzyme (81), thus indicating that the amino acid sequence is a more decisive factor than zoologic relationship. This finding implies that

Table 3. Immunologic Relationship among Several Bird Lysozymes[a]

Species	Number of amino acid differences from hen lysozyme	Extent of cross reactivity
Hen	0	100
Bobwhite quail	4	90–95
Japanese quail	6	90–92
Guinea fowl	10	71
Turkey	11	74–94
Duck A	22	66

[a] According to Ref. 80.

there is a correlation between the degree of antigenic difference and the degree of sequence difference among bird lysozymes (82).

In another study, which employed the modified bacteriophage technique (60,61), gradual similarity between the various bird lysozymes was likewise observed, in the order expected from animo acid sequence analyses (83). This sensitive technique could readily differentiate between two duck lysozymes that are on the same evolutionary level and differ by only limited sequence replacements, confined mainly to defined areas of the molecule (84). Furthermore, in this study, a small extent of cross reaction was observed also between the bird lysozymes and human lysozyme, obtained from chronic lymphatic leukemia patients (83,85), which indicates some similarity even at this distance on the evolutionary scale.

A comparative study was also performed on a series of primate lysozymes. In this case, only the amino acid sequence of human lysozyme has been determined (86,87); little is known about the lysozymes of other primates, except that tears and milk are rich in them (88). In the immunochemical study, antisera were produced against highly purified human and baboon lysozymes and were tested for reactivity with the lysozymes of 19 primate species (15). Microcomplement fixation tests revealed no antigenic differences between the lysozymes of man and chimpanzee. But, major differences were detected between human lysozyme and those of gorilla, Old World monkeys, New World monkeys, and prosimians. These results do not permit construction of an accurate phylogenetic tree of the primates, but they do permit estimation of the average rate of lysozyme evolution. Lysozyme evolution appears to have proceeded very rapidly in primates.

7.2 Structural Aspects

Upon immunologic comparison of lysozymes of various species, the observed differences are gradual and are more or less in accordance with the phylogenetic distance between the species and with the resemblance in amino acid sequences. When the phylogenetic distance is large, such as when hen and human lysozymes are compared, one can hardly observe any cross reaction (83,89); on the other hand, between various bird lysozymes, there are extensive cross reactions (78).

It has already been mentioned that antigenic specificity of globular proteins is dependent mainly on conformational determinants. Accordingly, the amino acid replacements that preserve the conformation of a determinant produce little alteration in antigenicity, whereas those that alter the conformation produce gross antigenic changes. The immunologic relationships among bird lysozymes, which show close similarity, indicate that the amino acid replacements that have survived in naturally occurring bird lysozymes are usually small, and their effect on conformational alteration is also small. Nevertheless, immunologic studies were able to detect differences and to assign them to particular regions or structural features of the molecule. When the immunologic reagents used include antibodies to whole lysozyme and antibodies to specific regions (such as the loop), it is possible to obtain selective information about the differences in defined regions of the molecule. For example, guinea hen lysozyme is a four times less efficient inhibitor than hen lysozyme for the interaction between lysozyme and antibodies to the intact molecule, whereas both have the same inhibitory capacity for the loop-antiloop system (83). In contrast, the two duck lysozymes, II and III, are indistinguishable in their inhibition of the lysozyme system but differ drastically in inhibiting the loop system; only duck lysozyme II is inhibitory. This finding is in agreement with the recognized replacement of the glycine residue in position 71 of the sequence of hen and duck lysozyme II with an arginine residue in duck lysozyme III (84). This introduction of a new basic amino acid might explain the immunologic difference.

In another study of a similar type (90), it was demonstrated that bobwhite quail lysozyme was as efficient as hen lysozyme in inhibition of the lysozyme system but was much less reactive with antiloop antibodies. Turkey lysozyme, on the other hand, was similar to hen lysozyme in its behavior with antiloop antibodies but different in its reactivity with antilysozyme. Each of these two lysozymes differs from hen lysozyme in the loop region by a single replacement, lysine for arginine, but in a different position (position 68 in the quail and position 73 in the turkey). The above findings are, therefore, indicative of a stronger influence of the arginine in position 68 than that in position 73 on the antigenic properties of this conformation-dependent region. This finding is in accord with x-ray crystallography data that indicate that Arg_{68} participates much more than Arg_{73} in hydrogen bonding to other residues in lysozyme (14).

In addition to the above-mentioned difference in position 68, which is within the loop region, bobwhite quail lysozyme differs from hen lysozyme in only three other positions, residues 40, 55, and 91, which are elsewhere in the protein (82). It has been observed that rabbit antihen lysozyme sera that lack antiloop antibodies are still capable of distinguishing between these two lysozymes in the complement fixation test (91). Because the residues at positions 40, 55, and 91 are fully buried in the interior of chicken lysozyme, it is possible to conclude that these replacements alter the antigenic properties of lysozyme by changing its conformation. It can thus be concluded that immunologic comparison of evolutionarily related proteins can assist in the elucidation of the role played by various molecular parameters, affected by amino acid replacements, in the local conformation of particular antigenic determinants in the molecule.

It has been mentioned earlier that between phylogenetically remote lysozymes, such as those of hen and human (which exhibit over 50% differences in their sequences), the extent of cross reaction between the native molecules is rather small. A very interesting observation was, therefore, the marked cross reactivity between their open-chain polypeptide chains, obtained by reduction and carboxymethylation (89). This phenomenon will be interpreted in detail in the next section, when discussing similar findings observed in the comparison of lysozyme and α-lactalbumin.

8. RELATIONSHIP BETWEEN LYSOZYME AND α-LACTALBUMIN

In connection with studies on the evolution of lysozyme, it was of interest to study the relationship between hen egg-white lysozyme and bovine α-lactalbumin. These two proteins show a striking similarity in their amino acid sequence; as shown in Fig. 5, 49 of 129 amino acid residues are identical, including an identity in the position of their disulfide bridges (87,92,93). On this basis, it has been suggested that they might be functionally related (92) and might also exhibit structural similarities (94). With respect to functional similarity, α-lactalbumin has been found to be identical to the B protein, one of the two naturally occurring subunits of lactose synthetase (95), although devoid of any catalytic activity by itself. In a way, the enzymatic reaction of lactose synthetase may be considered to be reversely related to the hydrolytic reaction catalyzed by

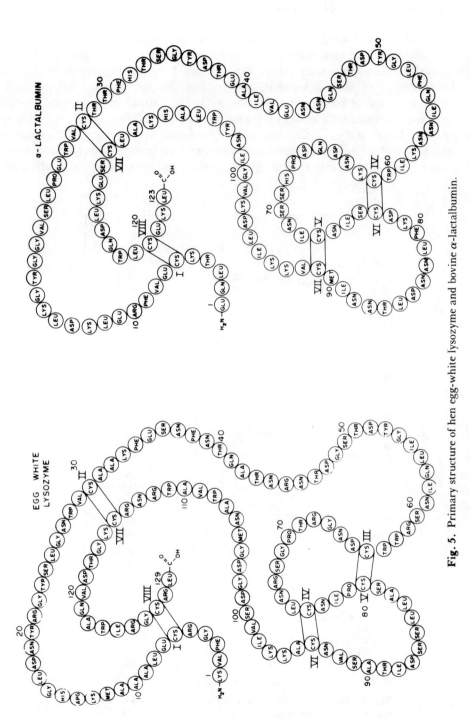

Fig. 5. Primary structure of hen egg-white lysozyme and bovine α-lactalbumin.

lysozyme. It was, therefore, postulated that the two proteins are evolutionarily related; that is, they may have been derived from a common ancestor gene by gene duplication (95), and this relationship might explain the structural homology between them.

Measurements of various physical properties of the two proteins did not lead, however, to a unified concept of the similarity in their structure (97). Also, because x-ray crystallographic data on α-lactalbumin are not yet available, it is impossible to determine whether there is a similarity in their conformation. An attempt was made to build a wire skeletal model of α-lactalbumin, based on the known main chain conformation of lysozyme (98). Comparison of the two models, and the readiness with which the

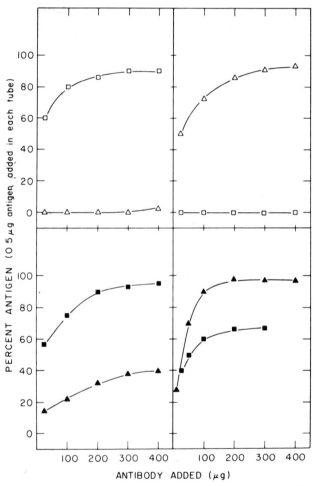

Fig. 6. Antigen-binding capacity of native and open-chain hen egg-white lysozyme and bovine α-lactalbumin by isolated antibodies against native lysozyme (*upper left*), native lactalbumin (*upper right*), reduced carboxymethylated lysozyme (*lower left*), and reduced carboxymethylated lactalbumin (*lower right*). The antigens are native lysozyme, $\square$; native lactalbumin, $\triangle$; reduced carboxymethylated lysozyme, $\blacksquare$; reduced carboxymethylated lactalbumin $\blacktriangle$.

changes in amino acid side chains could be accomodated in the proposed α-lactalbumin model, led to the conclusion that the two proteins may be structurally related, but this conclusion is merely a hypothesis.

In immunologic studies performed as a probe for the detection of a conformational relationship, antisera produced against either lysozyme or α-lactalbumin were tested for immunologic interactions with the two antigens by a variety of sensitive assays, such as antigen-binding capacity, passive cutaneous anaphylaxis, microcomplement fixation, passive hemagglutination, and phage inactivation. The antisera to each of the proteins, raised either in goats or in rabbits, exhibited a strong positive reaction with the homologous protein antigen but showed no cross reaction with the heterologous antigen (99).

On the other hand, when the unfolded peptide chains of lysozyme and α-lactalbumin (obtained by complete reduction and carboxymethylation) were used for immunization (100), each of the resultant antisera showed an appreciable extent of immunologic cross reaction that could be measured by various techniques, including the precipitin test and antigen-binding capacities (Fig. 6). These findings lead to several conclusions: (a) Antibodies to the native conformation are apparently more specific than antibodies raised against the unfolded peptide chain and fail to recognize similarities in amino acid sequence per se. (b) Lysozyme and α-lactalbumin, which are obviously related, differ in their "hydrophilic peripheries" that are exposed to the surrounding medium, namely, those areas that may lead to antibodies of similar specificities. They might still be similar in their three-dimensional overall conformation, which is dictated by the "internal" residues of the molecule and is the feature detected by x-ray crystallography.

It is noteworthy that, as mentioned in the preceding section, a similar phenomenon exists in the immunologic relationship between hen egg-white and human lysozymes. In this case, the degree of sequence homology is also high: 77 of 129 amino acid residues are in identical positions (101). When the immunologic resemblance of human leukemia and hen egg-white lysozymes was studied by complement fixation, no cross reaction could be detected between the native proteins (89), despite the evidence that they are quite similar in three-dimensional structure (102). In contrast, when the reduced carboxymethylated derivatives of these lysozymes were examined by the same immunologic techniques, each of them reacted with both antisera. These results indicate that immunologic studies with unfolded proteins may reveal sequence homologies that are undetected when the native molecules are examined, but this again does not rule out the possibility that the proteins are also conformationally related.

9. CELLULAR RECOGNITION AND HUMORAL RESPONSE TO CONFORMATIONAL AND SEQUENTIAL DETERMINANTS

All of the data and findings discussed so far argue strongly in support of the view that protein antigenic determinants that interact with humoral antibodies usually express conformational specificity. This relationship does not appear to hold, however, for the determinants that mediate cellular immunity. Several years ago, it was suggested that there are possible differences in both specificity and manifestation of cell-mediated response and antibody production (103). This assumption was borne out by the results obtained for several systems, which suggest that the essential molecular requirements (size and conformation) for eliciting a cellular immune response are less stringent that those needed for eliciting the formation of circulating antibodies. Thus, protein fragments that fail to stimulate detectable circulating antibody formation may, nevertheless, elicit cellular immunity. The best documented examples for this phenomenon are provided by a fragment of the basic protein of myelin (104) and a low-molecular-weight (< 4000) peptic peptide of flagellin (105). In both cases, the peptides induced cell-mediated immunity, as manifested by the delayed hypersensitivity reaction to the parent protein, in preference to circulating antibodies, and failed to react with an antiserum against the whole protein.

Similar observations were reported for high-molecular-weight molecules, namely, intact proteins. One example for this phenomenon is the cross reaction between various collagens. A comparison was made between the collagen of *Ascaris lumbricoides* and several vertebrate collagens. Physicochemical measurements have demonstrated that the worm collagen possesses a radically different structure than that of the vertebrate collagens (106). In parallel, it also appears to have completely distinct antigenic features and does not show any humoral cross reactivity with the vertebrate collagens (107). However, when tested for cellular response, both *in vivo* (delayed hypersensitivity) or *in vitro* (MIF), a marked extent of cross reactivity could be demonstrated between *Ascaris* and human collagens (108). This finding by itself is not sufficient evidence for either evolutionary relatedness or structural similarity, but it does indicate the relationship between the two proteins.

A similar phenomenon was observed concerning the relationship between hen egg-white lysozyme and bovine α-lactalbumin. As was discussed in the previous section, these two proteins in their native state are completely noncross reactive at the humoral antibody level. However, as has been reported recently, they exhibit a definite cross reaction at the cellular level. Thus, by use of both *in vivo* (delayed hypersensitivity) and *in vitro* (lymphocyte transformation) methods, a marked cross reactivity was observed between the two proteins (109), as demonstrated in Table 4. This phenomenon serves as another indication for the relationship between them.

That these immunologic events are related to conformational parameters is evident from the observation that similar phenomena exist on the level of the cross reaction between native and unfolded proteins, such as the case of lysozyme. As mentioned earlier (Sec. 3), the unfolded (reduced and carboxymethylated) lysozyme failed to cross react with antiserum to the native enzyme, and, vice versa, there was no cross reactivity between native lysozyme and antiserum to the unfolded chain. In contrast, extensive cross reactivity between native and unfolded lysozymes was detected on the cellular level (36). This reaction was manifested both *in vivo* by marked delayed hypersensitivity reaction (Table 4) and *in vitro* by the inhibition of capillary macrophage migration (MIF technique).

It appears that all of the observations reported here have a similar molecular basis, which can lead to the following conclusion: Whereas the specificity of humoral antibodies is strict and they recognize, and react with, mainly conformational antigenic determinants, the determinants that mediate cellular immunity are less conformation dependent. In other words, the conformational integrity of a protein molecule, though not crucial for interaction with cell-bound antibodies, is essential for eliciting humoral antibodies and reacting with them. Apart from emphasizing the different order of specificity recognized by humoral and cell-bound antibodies, the above results suggest that cell-mediated immunity represents a more informative parameter than humoral immunity for the study of amino acid sequence of phylogenetic relationships, especially

Table 4. Cross Reaction between Lysozyme and α-Lactalbumin or Unfolded Lysozyme in Delayed Hypersensitivity Test[a]

| | Skin reaction | | |
| | Sensitizing antigen | | |
Test material	Lysozyme	RCM[b] Lysozyme	α-Lactalbumin
Lysozyme	17.3 (3/3)	~12 (15/20)	7.3 (2/3)
RCM lysozyme	12 (7/8)	~12 (15/20)	nd[c]
α-Lactalbumin	7.0 (3/3)	nd	16.7 (3/3)
Human serum albumin (control)	2.7	<4	2.3
Phosphate-buffered saline	2.0	<4	2.0

[a] According to Refs. 36 and 109. The numbers denote the mean values of the diameter (mm) of the skin reaction, whereas the figures in parentheses denote the number of reacting guinea pigs in each experiment; nd, not determined.

[b] Reduced and carboxymethylated.

[c] Not determined.

between distantly related proteins (69,108). On the other hand, humoral antibodies are more useful for the study of structural and conformational relationships of proteins. In the particular case of lysozyme, the cell-mediated immune responses and the cross reactivities at the cellular level emphasize the relationship between lysozyme and related molecules, on the one hand, and the role of its primary structure in its immunochemical characterization, on the other hand.

10. IMMUNOGENETICS OF LYSOZYME

One of the most important aspects of immunology that has been very extensively studied in recent years is the genetic control of the immune response (110). The problem was approached mainly with synthetic polypeptides that form immunogens of relatively simple structure. These studies led to the conclusion that the antibody responses of inbred mouse strains are quantitative traits, under a dominant determinant-specific type of genetic control (111,112). Such studies were unapproachable by the use of natural multiple-determinant antigens, such as proteins, because the simultaneous production of antibodies of differing specificities complicated the study of the processes involved. One may very well envisage a situation where the specificity of antibodies formed in various inbred strains against a particular protein may vary, due to different antigenic determinants, but this effect could escape notice, because the overall amount of anti-bodies against the whole protein molecule might still be similar. In effect, this phenomenon was observed even in the case of a simpler synthetic antigen (multichain polyproline to which peptides of phenylalanine and glutamic acid were attached). Two inbred strains immunized with this antigen responded equally well to the whole polypeptide, but one of them produced antibodies directed mainly toward the side-chain peptide portion of the molecule, whereas the antisera of the second strain reacted primarily with the backbone moiety of the immunogen (21). It appears, therefore, that in this case, two different genetic controls are operating for two determinants of the same molecule. With multideterminant protein antigen, the situation is expected to be much more complex.

By use of the lysozyme-loop system, advantage was taken of the feasibility of eliciting antibodies reactive specifically with a unique region of this native protein. It was thus possible to analyze the immune response of various inbred mouse strains to the loop region, as compared to their response to other portions of the molecule (113,114). The results demonstrated that most of the mouse strains tested responded well when immunized with lysozyme, but only some strains produced antibodies when injected with the loop-A—L conjugate or with a similar conjugate that contained the loop attached to multichain polyproline (loop-Pro—L). The strains that did not respond to these conjugates were also unable to elicit antibodies specific toward the loop region when injected with intact lysozyme. Analysis of the immune response in the F_1 hybrid between low and high responders (showing intermediate response), and their back-crosses with the parental strains, demonstrated that the antibody production to the loop is, indeed, genetically controlled, by a unigenic dominant trait, which is not linked to the major histocompatibility locus H-2 (Fig. 7).

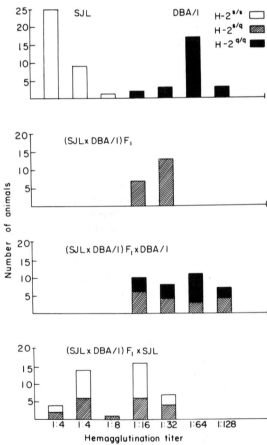

Fig. 7. The immune response of mice immunized with loop-A—L and titrated with lysozyme-coated sheep red blood cells by use of the passive microhemagglutination assay. Animals carrying the H-2^s allele donated by the SJL mice (□), animals carrying the H-2^q allele donated by the DBA/1 mice (■), animals carrying the H-2$^{s/q}$ alleles (▨).

The cellular aspects of the immune response gene that regulates the immune potential to the loop were also studied, by observing the effect of syngeneic and allogeneic thymus and bone marrow cell transfers in irradiated recipients immunized with loop-A—L, loop-Pro—L, or lysozyme (115). The results obtained from these cell transfer experiments indicated that whereas both thymus and marrow cells show a defect when both determinant and carrier are involved, the defect concerning the loop determinant per se was reflected exclusively in the marrow cells of the low-responder mice (116).

These experiments provide evidence that the genetic control of the immune response at the level of antigenic determinants holds not only for synthetic antigens, on which most of the studies were performed, but also for native proteins.

11. SUMMARY AND CONCLUDING REMARKS

In the last few years, we have witnessed an increased number of immunologic studies on enzymes that have demonstrated the fruitfulness of such investigations with respect to the many facets of enzyme research (117). With an enzyme such as lysozyme, whose three-dimensional structure has been elucidated in detail, the immunochemical analysis has, on the one hand, corroborated findings that had been obtained by other methods, but, on the other hand, in some cases it has offered a solution to problems to which ambiguous answers had been obtained by different procedures.

By use of the various approaches mentioned in Sec. 2, namely, chemical modification, fragmentation, and physical denaturation, it was possible to characterize the immunochemical properties of the molecule. Thus, on the one hand, the crucial role of structural conformation and its imprint on immunologic specificity was established, and, on the other hand, several defined regions of the lysozyme molecule were recognized as contributing to its antigenic reactivity. An interesting point is that although the total population of antibodies elicited by lysozyme is inhibitory for the catalytic activity of the enzyme, this capacity is the property of only a small fraction of the antibodies, whereas the rest are directed toward other regions of the molecule that are irrelevant to the catalytic site. Two such independent regions were clearly defined, the region confined in the sequence 57–107 in the protein and the N-C terminal peptide. Both regions contain disulfide bonds for stabilization of their structure and seem to occupy "corners" in the three-dimensional structure of the molecule.

A smaller and more defined antigenic region of the molecule, which is antigenic also when isolated after fragmentation, is the loop fragment, whose structure is also stabilized by an intrachain disulfide bond. This fragment was shown to maintain, in the isolated form, the same conformation that it assumes in the intact molecule. Antibodies elicited by this fragment, or by conjugates that contain it, are specific to the conformation-dependent determinant present in the native protein. Moreover, synthetic derivatives and analogs of the loop fragment are also antigenic, and changes in the positions that are crucial for its spatial conformation impair or abolish the antigenic reactivity.

As discussed in the last sections of this chapter, the availability of this system has made possible the study, on a molecular level, of several biologic aspects, such as the genetic control of the immune response elicited by a multideterminant protein molecule, the biochemical evolution both of isofunctional enzymes (phylogenetic relationships among various lysozymes) and of chemically homologous proteins that do not share a biologic function (lysozyme and α-lactalbumin), and the relationship between cellular recognition and humoral immune response of native proteins, as a corollary to their molecular properties.

The continuation of this type of study should be helpful not only in the mapping of antigenic structures of enzymes or proteins in general, and in elucidating the detailed features of their antigenic determinants, but also in the approach toward several aspects of biochemical research. In this connection, the high sensitivity of the immunologic techniques, on the one hand, and the selective specificity of the immunologic reactions, on the other, should serve as a powerful tool for molecular analysis.

REFERENCES

1. K. Landsteiner, *The Specificity of Serological Reactions,* Harvard University Press, Cambridge, Mass., 1945.
2. M. Sela, *Advan. Immunol.,* **5,** 29 (1966).
3. W. C. Boyd, *Fundamentals of Immunology,* Interscience, New York, 1962.
4. D. Pressman and A. L. Grossberg, *The Structural Basis of Antibody Specificity,* W. A. Benjamin Inc., New York, 1968.
5. M. J. Crumpton, in B. Cinader, Ed., *Antibodies to Biologically Active Molecules,* Pergamon Press, Oxford, 1967, p. 61.
6. M. Z. Atassi and B. J. Saplin, *Biochemistry, 7,* 688 (1968).
7. C. Lapresle and J. Darieux, *Ann. Inst. Pasteur, 92,* 62 (1957).
8. J. J. Cebra, *J. Immunol.,* **86,** 205 (1961).
9. E. Benjamini, J. D. Young, M. Shimizu, and C. Y. Leung, *Biochemistry, 3,* 1115 (1964).
10. M. Z. Atassi, *Immunochemistry,* 12, 423 (1975).
11. R. Arnon, *Curr. Topics Microbiol. Immunol., 54,* 47 (1971).
12. D. C. Phillips, *Sci. Amer.* 215, 78 (1966).
13. C. C. F. Blake, D. F. Koenig, G. A. Mair, A. C. T. North, D. C. Phillips, and V. R. Sarma, *Nature (London),* **206,** 757 (1965).
14. D. C. Phillips, *Proc. Nat. Acad. Sci. U.S.A., 57,* 484 (1967).
15. A. C. Wilson and E. M. Prager, in E. F. Osserman, R. E. Canfield, and S. Beychok, Eds., *Lysozyme,* Academic Press, New York, 1974, p. 127.
16. M. Z. Atassi, *Immunochemistry, 12,* 423 (1975).
17. A. F. S. A. Habeeb and M. Z. Atassi, *Immunochemistry, 6,* 555 (1969).
18. A. D. Strosberg and L. Kanarek, *FEBS Lett., 5,* 324 (1969).
19. B. Bonavida, Dissertation Thesis, University Microfilms, Ann Arbor, Mich., 1968.
20. M. Z. Atassi and A. F. S. A. Habeeb, *Biochemistry, 8,* 1385 (1969).
21. A. F. S. A. Habeeb, *Arch. Biochem. Biophys., 121,* 652 (1967).
22. A. F. S. A. Habeeb and M. Z. Atassi, *Immunochemistry, 8,* 1047 (1971).
23. A. D. Strosberg and L. Kanarek, *Arch. Int. Physiol. Biochem., 76,* 949 (1968).
24. A. F. S. A. Habeeb and M. Z. Atassi, *Immunochemistry, 9,* 4939 (1970).
25. M. Z. Atassi, M. C. Rosemblatt, and A. F. S. A. Habeeb, *Immunochemistry, 11,* 495 (1974).
26. M. Z. Atassi and M. C. Rosemblatt, *J. Biol. Chem., 249,* 4802 (1974).
27. M. Z. Atassi, A. M. Suliman, and A. F. S. A. Habeeb, *Immunochemistry, 9,* 907 (1972).
28. M. J. Crumpton and P. A. Small, *J. Mol. Biol., 26,* 143 (1967).
29. H. Taniuchi and C. B. Anfinsen, *J. Biol. Chem., 246,* 2291 (1971).
30. B. Schechter, A. Conway-Jacobs, and M. Sela, *Eur. J. Biochem., 20,* 321 (1971).
31. M. Sela, B. Schechter, I. Schechter, and F. Borek, *Cold Spring Harbor Symp. Quant. Biol.,* **32,** 537 (1967).

32. R. E. Dickerson, in H. Neurath, Ed., *The Proteins*, 2nd edit., Vol. 2, Academic Press, New York, 1964, p. 634.
33. W. A. Hendrickson and W. E. Love, *Nature [New Biol.]*, **232**, 197 (1971).
34. J. Gerwing and K. Thompson, *Biochemistry, 7*, 3888 (1968).
35. J. D. Young and C. Y. Leung, *Biochemistry, 9*, 2755 (1970).
36. K. Thompson, M. Harris, E. Benjamini, G. Mitchell, and M. Noble, *Nature [New Biol.]*, **238**, 20 (1972).
37. L. Ching-Li and M. Z. Atassi, *Biochemistry, 12*, 2690 (1973).
38. R. Lumry and H. Eyring, *J. Phys. Chem., 58*, 110 (1954).
39. M. Sela, F. H. White, Jr., and C. B. Anfinsen, *Science, 125*, 691 (1957).
40. F. H. White, Jr. and C. B. Anfinsen, *Ann. N.Y. Acad. Sci., 81*, 515 (1959).
41. C. B. Anfinsen, *J. Polymer Sci., 49*, 31 (1961).
42. C. B. Anfinsen, in M. Sela, Ed., *New Perspectives in Biology*, Elsevier, Amsterdam, 1964, p. 42.
43. F. H. White, Jr., *J. Biol. Chem., 235*, 383 (1960).
44. R. P. Singhal and M. Z. Atassi, *Biochemistry, 9*, 4252 (1970).
45. S. Shinka, M. Imanishi, O. Kuwahara, H. Fujio, and T. Amano, *Biken J., 5*, 181 (1962).
46. R. E. Canfield and A. K. Liu, *J. Biol. Chem., 240*, 1997 (1965).
47. S. Shimka, M. Imanishi, N. Miyagawa, T. Amano, M. Inouye, and A. Tsugita, *Biken J., 10*, 89 (1967).
48. R. Arnon, *Eur. J. Biochem., 5*, 583 (1968).
49. H. Fujio, M. Imanishi, K. Nishioka, and T. Amano, *Biken J., 11*, 207 (1968).
50. E. Maron, Ph.D. Thesis, Feinberg Graduate School, Rehovot, Israel, 1971.
51. H. Fujio, M. Imanishi, K. Nishioka, and T. Amano, *Biken J., 11*, 219 (1968).
52. B. Cinader, *Ann. N.Y. Acad. Sci., 103*, 495 (1963).
53. R. von Fallenberg and L. Levine, *Immunochemistry, 4*, 363 (1967).
54. M. Imanishi, N. Miyagawa, H. Fujio, and T. Amano, *Biken J., 12*, 85 (1969).
55. S. Nobuo, H. Fujio, and T. Amano, *Biken J., 15*, 135 (1972).
56. M. Z. Atassi, A. F. S. A. Habeeb, and K. Ando, *Biochim. Biophys. Acta, 303*, 203 (1973).
57. H. Fujio, R. E. Martin, Y.-M. Ha, N. Sakato, and T. Amano, *Biken J., 17*, 73 (1974).
58. R. Arnon and M. Sela, *Proc. Nat. Acad. Sci. USA, 62*, 163 (1969).
59. E. Maron, C. Shiozawa, R. Arnon, and M. Sela, *Biochemistry, 10*, 763 (1971).
60. J. Haimovich, E. Hurwitz, N. Novik, and M. Sela, *Biochim. Biophys. Acta, 207*, 115 (1970).
61. J. Haimovich, E. Hurwitz, N. Novik, and M. Sela, *Biochim. Biophys. Acta, 207*, 125 (1970).
62. I. Pecht, E. Maron, R. Arnon, and M. Sela, *J. Biochem., 19*, 368 (1971).
63. R. B. Merrifield, *Science, 150*, 178 (1965).
64. R. Arnon, E. Maron, M. Sela, and C. B. Anfinsen, *Proc. Nat. Acad. Sci. USA, 68*, 1450 (1971).
65. E. Teicher, E. Maron, and R. Arnon, *Israel J. Med. Sci., 8*, 631 (1972).
66. E. Teicher, E. Maron, and R. Arnon, *Immunochemistry, 10*, 265 (1973).
67. R. Arnon, in E. R. Blout, F. A. Bovey, M. Goodman, and N. Lotan, Eds., *Peptides, Polypeptides and Proteins*, John Wiley and Sons, New York, 1974, p. 538.
68. E. Benjamini, D. Michaeli, and J. D. Young, *Curr. Topics Microbiol. Immunol., 52*, 85 (1972).
69. M. J. Crumpton, in M. Sela, Ed., *The Antigens*, Vol. II, Adademic Press, New York, 1974, p. 1.
70. J. G. Levy, D. Hull, B. Kelly, D. G. Kilborn, and R. M. Teather, *Cell Immunol., 5*, 87 (1972).
71. B. Geiger and R. Arnon, *Eur. J. Immunol., 4*, 632 (1974).
72. I. Pilz, O. Kratky, A. Licht, and M. Sela, *Biochemistry, 12*, 4998 (1973).
73. I. Pilz, O. Kratky, A. Licht, and M. Sela, *Biochemistry, 14*, 1326 (1975).
74. J. Schlessinger, I. Z. Steinberg, D. Givol, J. Hochman, and I. Pecht, *Proc. Nat. Acad. Sci. USA, 72*, 2775 (1975).
75. I. Pecht, B. Ehrenberg, E. Calef, and R. Arnon, unpublished data.
76. R. C. Valentine and N. M. Green, *J. Mol. Biol., 27*, 615 (1967).

77. N. E. Hyslop, Jr., R. R. Dourmashkin, N. M. Green, and R. R. Porter, *J. Exp. Med.,* **131,** 783 (1970).

77a. V. N. Schumaker, G. Green, and R. L. Wilder, *Immunochemistry,* **10,** 521 (1973).

78. N. Arnheim, Jr. and A. C. Wilson, *J. Biol. Chem.,* **242,** 3951 (1967).

79. E. M. Prager and A. C. Wilson, *J. Biol. Chem.,* **246,** 5978 (1971).

80. E. M. Prager and A. C. Wilson, *J. Biol. Chem.,* **246,** 7010 (1971).

81. N. Arnheim, E. M. Prager, and A. C. Wilson, *J. Biol. Chem.,* **244,** 2085 (1969).

82. E. M. Prager, N. Arnheim, G. A. Mross, and A. C. Wilson, *J. Biol. Chem.,* **247,** 2905 (1972).

83. E. Maron, R. Arnon, M. Sela, J.-P. Perrin, and P. Jollés, *Biochim. Biophys. Acta,* **214,** 222 (1970).

84. J. Jollés, B. Nieman, J. Hermann, and P. Jollés, *Eur. J. Biochem.,* **1,** 344 (1967).

85. A. Miller, B. Bonavida, S. A. Stratton, and E. Sercarz, *Biochim. Biophys. Acta,* **243,** 520 (1971).

86. R. E. Canfield, S. Kammerman, J. H. Sobel, and F. J. Morgan, *Nature [New Biol.],* **232,** 16 (1971).

87. J. Jollés and P. Jollés, *Helv. Chim. Acta,* **54,** 2668 (1971).

88. D. H. Buss, *Biochim. Biophys. Acta,* **236,** 587 (1971).

89. N. Arnheim, J. Sobel, and R. E. Canfield, *J. Mol. Biol.,* **61,** 237 (1971).

90. M. Fainaru, A. C. Wilson, and R. Arnon, *J. Mol. Biol.,* **84,** 635 (1974).

91. E. M. Prager, M. Fainaru, A. C. Wilson, and R. Arnon, *Immunochemistry,* **11,** 153 (1974).

92. K. Brew, T. C. Vanaman, and R. L. Hill, *J. Biol. Chem.,* **242,** 3747 (1967).

93. R. E. Canfield, F. J. Morgan, S. Kammermann, J. J. Bell, and G. M. Agosto, in E. B. Astwood, E., *Recent Progress in Hormone Research,* Vol. 27, Academic Press, New York, 1971, p. 121.

94. P. N. Lewis and H. A. Scheraga, *Arch. Biochem. Biophys.,* **144,** 584 (1971).

95. U. Brodbeck, W. L. Denton, N. Tanahashi, and K. A. Ebner, *J. Biol. Chem.,* **242,** 1391 (1967).

96. R. L. Hill, K. Brew, T. C. Vanaman, I. P. Trayer, and P. Mattock, *Brookhaven Symp. Biol.,* **21,** I, 139 (1969).

97. A. M. Tamaburro, G. Jori, G. Vidali, A. Scatturin, and G. Saccomani, *Biochim. Biophys. Acta,* **263,** 704 (1972).

98. W. J. Browne, A. C. T. North, D. C. Phillips, K. Brew, T. C. Vanaman, and R. L. Hill, *J. Mol. Biol.,* **42,** 65 (1969).

99. R. Arnon and E. Maron, *J. Mol. Biol.,* **51,** 703 (1970).

100. R. Arnon and E. Maron, *J. Mol. Biol.,* **61,** 225 (1971).

101. J. B. C. Findlay and K. Brew, *Eur. J. Biochem.,* **27,** 65 (1972).

102. C. C. F. Blake and I. D. A. Swan, *Nature (London),* **232,** 12 (1971).

103. G. Senyk, E. B. Williams, D. E. Nitecki, and J. W. Goodman, *J. Exp. Med.,* **133,** 1294 (1971).

104. V. A. Lennon, A. V. Wilks, and P. R. Carnegie, *J. Immunol.,* **105,** 1223 (1970).

105. L. B. Ichiki and C. R. Parish, *Immunochemistry,* **9,** 153 (1972).

106. O. W. McBride and W. F. Harrington, *Biochemistry,* **6,** 1484 (1967).

107. S. Fuchs and W. F. Harrington, *Biochim. Biophys. Acta,* **221,** 119 (1970).

108. D. Michaeli, A. Senyk, A. Maoz, and S. Fuchs, *J. Immunol.,* **109,** 103 (1972).

109. E. Maron, C. Webb, D. Teitelbaum, and R. Arnon, *Eur. J. Immunol.,* **2,** 294 (1972).

110. H. O. McDevitt and B. Benaceraff, *Advan. Immunol.,* **11,** 31 (1969).

111. P. Pinchuk and P. H. Maurer, in B. Cinader, Ed., *Regulation of the Antibody Response,* Charles C Thomas, Springfield, Ill., 1968, p. 67.

112. H. O. McDevitt and M. Sela, *J. Exp. Med.,* **122,** 517 (1965).

113. E. Mozes, E. Maron, R. Arnon, and M. Sela, *J. Immunol.,* **106,** 862 (1971).

114. E. Maron, H. I. Scher, E. Mozes, R. Arnon, and M. Sela, *J. Immunol.,* **111,** 101 (1973).

115. E. Mozes, G. M. Shearer, E. Maron, R. Arnon, and M. Sela, *J. Immunol.,* **111,** 1429 (1973).

116. M. Sela, *Harvey Lect.,* **67,** 213 (1973).

117. R. Arnon, in M. Sela, Ed., *The Antigens,* Vol. I, Academic Press, New York, 1973, p. 88.

CHAPTER 2

ENZYME/ANTISERUM INTERACTIONS OF β-LACTAMASES

M. H. RICHMOND

1. INTRODUCTION

The β-lactamases (penicillin amido hydrolases: E.C. 3.5.2.6.) and cephalosporin amido hydrolases (E.C. 3.5.2.8.) have been extensively studied from the point of view of their chemical nature and their physiologic relevance to the bacterial cells that produce them (1–7), but relatively little attention has been paid to their immunologic properties. What has been done has largely been as an aide to protein studies.

Despite this relative shortage of specific studies, there, nevertheless, exists a fair amount of information on the immunologic reaction of these enzymes scattered through papers primarily concerned with other things. This chapter will attempt to summarize this work and to accumulate the relevant references in one place. It also draws on a fair amount of unpublished material from various sources, notably from the Department of Molecular Biology at Edinburgh University, Edinburgh, Scotland. In addition, it attempts to assess the extent to which the penicillinase/antipenicillinase system might be used for basic studies on enzyme/antiserum interactions.

Before embarking on this topic, it is essential to review briefly what is known about the various types of bacterial β-lactamases and their relationships with one another. Nomenclature in this field is notoriously confused, and this means that appreciation of the immunologic data can be difficult, particularly for those entering the field from the outside.

2. THE ENZYMES

The β-lactamases are widely distributed throughout the microbial kingdom (1). Moreover, a wide range of molecular variants may be found within most bacterial species. Part of this complexity is because some of the enzymes are plasmid mediated and therefore found in any species accessible to the plasmid in question (2); in other cases, however, a range of closely related molecules may be found in various strains of a given species under conditions where the genes concerned have every appearance of being chromosomal (6). Some isolates may even express two distinct enzymes (2). At present, we are still far from certain as to the precise number of different β-lactamases to be found in bacteria, and a study of the literature in this area gives a clear impression of a rapidly evolving enzyme system with a large number of molecular variants present, with varying abundances, in the bacterial population.

2.1 β-Lactamases of Gram-positive Bacteria

2.1.1. Bacillus cereus

The enzymes from Bacillus cereus, Bacillus licheniformis, and Staphylococcus aureus were the first β-lactamases to be studied in detail, the enzymes from Bacillus spp. because their induction provided a good experimental system for studying this phenomenon (8), and staphylococcal penicillinase because of its clinical importance (9). In B. cereus, two main molecular variants of penicillinase were detected, one characterized by the enzyme synthesized by strain NRRL 569 (or its constitutive mutant, 569/H) and the other from strain 5/B (10, 11). Originally, it was thought that the enzyme from strain 5/B was a somewhat larger molecule than that from strain 569/

H. However, more recent studies (11a) suggest that these two molecules are really very similar in their molecular weight and primary polypeptide sequences. In fact, these two variants are so similar that the difference between them was first detected by immunologic means (12).

The substrate profiles of the penicillinases synthesized by *B. cereus* 569/H and 5/B are all but indistinguishable (Table 1), as are their overall amino acid compositions (Table 2). Unpublished experiments (12a) have elucidated the primary sequence of five large fragments, which account for about 237 of the 270 residues of the enzyme from strain 569/H. These fragments show very considerable relatedness to similar sections of the primary sequence of the enzyme from *B. licheniformis* and *S. aureus* (see later).

The penicillinase from *B. cereus* 569/H has a ragged amino terminus, and the structures of several variants have been elucidated (3). For example, the amino terminal sequence of four variants from a single preparation had the structure:

Lys-His-Lys-Asx-Glx-Ala-Thr-
His-Lys-Asx-Glx-Ala-Thr-
Lys-Asx-Glx-Ala-Thr-
Asx-Glx-Ala-Thr-

This range of structures may account for the electrophoretic inhomogeneity of many preparations of this enzyme.

The cephalosporinase activity found in the culture supernatants of induced cultures of *B. cereus* 569, 569/H, and 5/B is almost exclusively due to a distinct enzyme, which is now known as β-lactamase II. Its substrate profile is shown in Table 1. This enzyme is activated by Zn^{2+} ions and is much more thermostable than the original β-lactamase I. The molecular weight of β-lactamase II is much less than that of the type-I enzyme (22,800 versus 28,000), and its overall amino acid composition (Table 2) and what is known of its primary sequence (12a) show that it is a very different molecule (3,13–15). Moreover, its synthesis is specified by a gene distinct from that concerned with the sequence of β-lactamase I (16).

Further details of the chemical makeup of these enzymes should be sought in the summary by Thatcher (3).

Most of the studies on type-I β-lactamase have been performed either on purified enzyme or on supernatant fractions. However, this enzyme is not completely liberated

Table 1. Substrate profiles of *Bacillus cereus* β-lactamase

	β-Lactamase I		β-Lactamase II
Substrate	Strain 569/H	Strain 5/B	Strain 569/H
Benzylpenicillin	100[a]	100	100
Methicillin	4	7	120
Ampicillin	200	160	50
6-APA[b]	12	—	10
Cephalosporin C	0.1	—	14
Cephaloridine	1		18

[a] All values ore relative to a rate of hydrolysis of benzylpenicillin equal to 100. Data are from Refs. 1, 3, 6, 10, and 11.

[b] 6-Aminopenicillanic acid.

Table 2. Amino Acid Composition of *Bacillus cereus* β-lactamase[a]

| Amino acid | β-Lactamase I | | β-Lactamase II |
	Strain 569/H	Strain 5/B[b]	Strain 569/H
Asp	33	32	24
Thr	22	(9)[c]	15
Ser	13	(2)	10
Glu	25	25	18
Pro	10	6	5
Gly	21	19	19
Ala	30	29	11
Val	15	17	20
Cys	0	0	1
Met	4	5	2
Ile	22	20	11
Leu	19	17	23
Tyr	9	7	5
Phe	7	7	4
His	4	6	6
Lys	23	29	19
Arg	13	13	5
Trp	3	?	(2)
Approximate molecular weight	28,000	28,000	22,800

[a] Data from Refs. 1, 3, and 6

[b] The values for this enzyme have been corrected to a molecular weight of 30,000 rather than 35,000, originally thought to be an accurate estimate.

[c] The values in parentheses are uncertain because of the methods used.

from the surface of exponentially growing *B. cereus*. At any one time, about 10% of the enzyme is cell bound, and there is some evidence that this enzyme may represent a conformational variant of the typical type-I enzyme (17). Originally, this cell-bound material (the so-called γ-penicillinase) was claimed to be distinct from the extracellular (or α enzyme), but Citri and colleagues have shown that, both by alteration of pH and by urea treatment, they can effect a γ to α conversion (7,18,19). However, these results have been challenged, and the exact relationship between these two forms of the enzyme is still not fully elucidated. The main reason for mentioning them here is that they have to be taken into account when discussing immunologic studies on *B. cereus* β-lactamase (17).

2.1.2.   *Bacillus licheniformis*

Molecular variants of β-lactamase are also found among strains of *B. licheniformis*. Again, two main variants may be distinguished, those typified by strains 749 and 6346 (originally described as *Bacillus subtilis* 6346). As is the case with the enzymes from *B. cereus*, the β-lactamases from these two strains have been purified, and their molecular characteristics have been determined. Substrate profiles are given in Table 3, but for a fuller account, Thatcher's paper (4) should be consulted. Both enzymes have a molecular weight of about 28,000 and a very similar amino acid composition (Table 4). Indeed, they probably differ by only five residues in their primary sequences. Most of the complete sequence of the enzyme from strain 749 has been published (20).

Table 3. Substrate Profile of *Bacillus licheniformis* β-lactamase

	Strain 749	Strain 6346
Benzylpenicillin	100[a]	100
Methicillin	0.5	1
Ampicillin	68	12
6-APA[b]	5	13
Cephalosporin C	1	1
Cephaloridine	36	160

[a] All rates adjusted to the rate of hydrolysis of benzyl penicillin equal to 100. Data are from Refs. 1, 4, 6, and 32.
[b] 6-Aminopenicillanic acid.

In addition to these two main variants found among strains of *B. licheniformis*, individual isolates often produce several molecular variants. This is most simply seen by the appearance of several "isoenzyme" bands when pure preparations of these enzymes are subjected to electrophoresis or isoelectric focusing. The variants that have been examined have all lost sequences from the amino terminus of the molecule, presumably by proteolytic action against the enzyme after synthesis (4) (see Sec. 2.1.1).

The enzymes from *B. licheniformis* are like those of *S. aureus* in that they are only partially liberated from the cells during exponential growth. The cell-bound enzyme has been studied extensively for its physiologic role (21–24), but little is yet known about its precise molecular composition *in situ*, although a complex with phospholipids does seem to be involved at some stage. The cell-bound enzyme can be liberated into a form very

Table 4. Amino acid analyses of extracellular *Bacillus licheniformis* β-lactamase[a]

Amino acid	Strain 749	Strain 6346
Asp	37	36
Thr	21	21
Ser	11	10
Glu	27	27
Pro	11	11
Gly	15	16
Ala	26	26
Val	15	14
Cys	0	0
Met	5	6
Ile	14	14
Leu	27	27
Tyr	6	6
Phe	7	7
Trp	3	3
Lys	24	24
His	1	1
Arg	15	16
Amides	20	20
Approximate molecular weight	28,500	28,500

[a] Data are from Ref. 4.

Table 5. Substrate Profiles of *Staphylococcus aureus* β-lactamase[a]

	Enzyme variant			
Substrate	Type A	Type B	Type C	Type D
Benzylpenicillin	100	100	100	100
Methicillin	0	0	0	0
Ampicillin	150	150	150	150
6-APA[b]	1.0	2.0	1.0	—
Cephalosporin C	0.1	1.0	0.1	—
Cephaloridine	23	20	20	—

[a] All values have been adjusted relative to benzylpenicillin equal to 100. Data are from Refs. 1, 4, 6, and 32.
[b] 6-Aminopenicillanic acid.

similar to the extracellular enzyme by treatment of the cells with trypsin, chymotrypsin, or sodium dodecyl sulfate (4).

2.1.3   *Staphylococcus aureus*

The β-lactamase from this species is probably the most important of the enzymes from gram-positive organisms in view of its major impact on the efficacy of benzylpenicillin as a therapeutic agent. Currently, more than 70% of the strains of *S. aureus* isolated in hospitals are resistant to this antibiotic because of the production of one or another of the four types of β-lactamases expressed by isolates of this species.

Initially, only three molecular variants of this type of enzyme were recognized (25), but a fourth variant has recently been added (26). All are extremely similar in terms of their amino acid composition, and, indeed, the distinction between the enzymatic types is made by studying their reaction with specific antiserum (25,26). The four variants are known as A, B, C, and D, and their molecular properties are summarized in Tables 5 and 6. Further details may be found in the original papers (25–27) and in the review on staphylococcal penicillinase (5). Most strains of *S. aureus* liberate about 50% of their β-lactamase during exponential growth (27), and, like the enzyme in *B. licheniformis*, the cell-bound enzyme is assumed to be the metabolic precursor of the extracellular material. To date, all attempts to purify the cell-bound material have failed, because the enzyme is tightly bound to the cytoplasmic membrane.

2.2   β-Lactamases of Gram-negative Bacteria

The β-lactamases of gram-negative species differ from their gram-positive counterparts in several aspects. Whereas the former are usually predominantly extracellular (1), those from gram-negative bacteria are intracellular and commonly lie either within the bacterial periplasmic space or on the outer surface of the bacterial inner membrane (2,28). Furthermore, gram-negative species usually synthesize much less enzyme than do gram-positives, at least after the latter have been induced.

Among gram-positive species, types of β-lactamases are species specific, and several variants are usually found within a given species. Among gram-negatives, the situation is more complex. First, the fact that some β-lactamases are R-factor mediated ensures that "the same" enzyme may be found within a range of species or even of genera (2). On the other hand, when this is the situation, variants are encountered, as is the case with gram-positive bacteria. Even among the enzymes specified by R factors, variants may be found (29). All of these data present a rather complex situation. So far, only very few of this considerable number of enzymes have been investigated with antisera, largely because only a minority of β-lactamases have been purified to the point of allowing any quantitative immunologic studies to be performed with any confidence.

Recent publications on the β-lactamases from gram-negative species have listed 14 distinct types of enzyme (2), but subsequent unpublished studies suggest that there certainly are more (29a). There is as yet no really satisfactory means of classifying these enzymes, because, in general, very little is known of their detailed chemical properties. All of the existing classification systems have therefore tended to rely on substrate and inhibitor profiles, with some consequently arbitrary grouping of certain patterns emerging.

The classification used in this article is based on one described earlier (2). It allocates the enzymes to five main groups, according to the following criteria:

Group I: Enzymes predominantly active against cephalosporins but inhibited by carbenicillin.

Group II: Enzymes predominantly active against penicillin.

Table 6. Amino acid analyses of *Staphylococcus aureus* β-lactamase[a]

| | Enzyme variant | |
Amino acid	Type A	Type C
Asp	39	38
Thr	13	13
Ser	19	17
Glu	18	17
Pro	9	9
Gly	12	13
Ala	18	19
Val	16	18
Cys	0	0
Met	3	3
Ile	19	20
Leu	22	21
Tyr	13	13
Phe	7	6
His	2	2
Lys	43	42
Arg	4	4
Trp	0	0

[a] Data are from Refs. 20 and 43a. Type-B enzyme has been analyzed on small quantities of material and gave patterns broadly similar to those of types A and C. Type-D enzyme has not yet been analyzed for its amino acid composition.

Group III: Enzymes with approximately equal activity against both penicillins and cephalosporins but that are sensitive to cloxacillin inhibition.

Group IV: Enzymes with profiles similar to those in group III but that are resistant to cloxacillin and sensitive to sulfhydryl inhibitors.

Group V: Enzymes with profiles similar to those in group III but that will hydrolyze cloxacillin and are sensitive to sulfhydryl inhibitors.

Within these five groups, several distinct types may be recognized, and Table 7 summarizes the situation as it was in mid 1973. Recent studies, particularly among strains of *Proteus* sp., has revealed several additional enzyme types (29a).

2.3 β-Lactamases of Mycobacteria

Kasik and collaborators have studied the β-lactamases of Mycobacteria (30,31). Activity was detected in cultures of *Mycobacterium smegmatis, Mycobacterium fortuitum,* and *Mycobacterium phlei,* and studies on the substrate profiles of the enzymes suggested that these species made three distinct types of enzyme (30). To date, little information has yet been published about the detailed protein chemistry of these molecules; they are mentioned here only because they were utilized in one comparative immunologic study (31).

2.4 Chemical Relationships Among Enzymes of Various Species

Apart from the obvious point that all β-lactamases open the β-lactam bond, all of these enzymes also share a relative sensitivity to I_2/KI solution (7). This tendency is at its greatest among the enzymes from gram-negative species, but even the enzymes from *B. cereus, B. licheniformis,* and *S. aureus* are relatively much more sensitive to iodine solution than is the "average" enzyme. This common feature argues that all β-lactamases open the β-lactam bond in the same general way and suggests that a tyrosine residue may always be involved as a pivotal group in the active center, because this amino acid is the most susceptible to iodination. This possibility is strengthened by the discovery that the activity of the penicillinase from all gram-positive species so far investigated is inhibited by mild treatment with tetranitromethane (3,12a), a reagent known to react preferentially with protein-bound tyrosine. In *B. cereus* type-I β-lactamase, the sequence that surrounds the tetranitromethane-sensitive tyrosine residue is -Thr-Lys-Glu-Asp-Leu-Val-Asp-*Tyr*-Ser-Pro-Ile-Thr-Glu-, whereas in *B. licheniformis* β-lactamase, it is -Thr-Arg-Asp-Asp-Leu-Val-Asn-*Tyr*-Asn-Pro-Ile-Thr-Glu-, and in the enzyme from *S. aureus,* it is -Asn-Lys-Asp-Asp-Ile-Val-Ala-*Tyr*-Ser-Pro-Ile-Leu-Glu-.

Even apart from this evidence for a common way of hydrolyzing the β-lactam bond, sequence studies show considerable similarities between some β-lactamases with very different substrate specificities. Similarity, in some cases, therefore extends further than just the nature of the active center. The best studied examples of this phenomenon are the enzymes from *B. licheniformis* and *S. aureus* (20). Comparison of Tables 4 and 5 shows that the overall amino acid composition of these two enzymes is very different, and this difference is reinforced by the fact that the staphylococcal enzyme is a strongly basic protein (27), whereas the β-lactamase from *B. licheniformis* is mildly acidic (32). Yet, elucidation of the primary sequence of these two β-lactamases shows that at least 40% of the amino acid residues in these two molecules occupy "equivalent" positions (20).

Table 7. The overall classification of β-lactamases from gram-negative bacteria based on substrate profiles[a]

Enzyme class	Enzyme type	Benzylpenicillin	Ampicillin	Carbenicillin	Cloxacillin	Cephaloridine	Cephalexin
				Substrate profile			
I	a	100	0	0	nd[b]	8000	620
	b	100	0	0	nd	350	80
	c	100	150	nd	nd	2000	nd
	d	100	10	0	0	600	80
II	a	100	180	45	nd	10	0
	b	100	160	nd	0	10	0
III	a	100	180	10	0	140	5
IV	a	100	120	10	5	150	0
	b	100	125	45	20	50	5
	c	100	170	50	20	70	0
V	a	100	950	nd	200	120	nd
	b	100	300	nd	200	50	nd
	c	100	100	60	0	20	5
	d	100	180	80	0	40	5

[a] Data are from Ref. 2.
[b] Not determined.

Experiments by Thatcher (12a) have shown that at least 80% of the residues of β-lactamase type I from *B. cereus* are equivalent to residues in the primary sequence of the *B. licheniformis* enzyme. This means, of course, that there is considerable similarity between the primary sequences of all of the β-lactamases from gram-positive species that have been examined so far; therefore, although these enzymes seem to have evolved to serve rather different physiologic requirements, evidence, nevertheless, remains of a common evolutionary origin (33). Perhaps the moral to be drawn from this study is that similarity in amino acid composition may well indicate similar enzymes, but differences do not necessarily mean substantially different molecules.

Sequence studies on other β-lactamases are far less well advanced, with the possible exception of the type-IIIa enzyme, which is R-factor mediated in many strains of enteric bacteria (34). The types-I and -II β-lactamases from *B. cereus* seem to have very different primary sequences (12a). This difference was not unexpected, because β-lactamase II requires Zn^{2+} ions for activity, whereas enzyme I does not (13–15). There also seems to be little primary sequence similarity between *B. cereus* enzyme II, on the one hand, and the penicillinases from *B. licheniformis* and *S. aureus*, on the other, although the complete sequence of the *B. cereus* β-lactamase has not yet been completely determined.

Probably, underlying similarities also exist among the enzymes found in gram-negative species, despite very different substrate and inhibitor profiles. The overall amino acid analyses of purified samples of types Ia, Ib, IIIa, and IVb enzymes are available (Table 8), and the enzymes are certainly similar (35). Moreover, this similarity exists despite the fact that group-I enzymes are primarily basic cephalosporinases, which are

Table 8. Amino acid composition of four distinct β-lactamases from enteric bacteria[a]

Amino Acid	Amino acid composition (% by weight)			
	Enzyme type			
	Ia	III	IVa	Ib
Lys	9.1	8.5	9.4	8.2,
His	2.1	1.0	1.0	1.9
Arg	4.4	5.6	5.7	5.5
Cys	0	0	0.8	0
Asp	13.8	13.6	11.8	9.8
Thr	7.2	7.4	7.4	6.5
Ser	4.3	4.4	6.4	4.9
Glu	13.7	15.2	12.2	14.6
Pro	5.5	5.7	3.5	5.5
Gly	3.2	3.4	4.6	4.5
Ala	5.0	6.3	6.2	6.6
Val	4.9	5.8	5.8	5.6
Met	1.9	1.9	1.9	1.9
Ile	4.0	4.2	4.1	5.6
Leu	8.8	10.0	9.1	8.8
Tyr	5.8	2.4	4.8	5.8
Phe	8.4	4.4	5.6	4.3
Trp	?	?	?	?

[a] Data are from Refs. 2 and 35.

inhibited by carbenicillin, whereas group-III enzymes are mildly acidic proteins with a broad spectrum of activity against many penicillins and cephalosporins, including carbenicillin (2,36). Perhaps the only β-lactamase that is clearly different from most enzymes from gram-negative species is the one specified by R1818. Here, the molecular weight of 44,000 is sharply different from those of all of the other enzymes, which have values closer to 25,000 (37,38).

Other than the sensitivity to I_2/KI solution (7), no molecular similarity between the enzymes from gram-positive and gram-negative species has yet been detected. In fact, all of the available evidence points away from a close similarity, and the enzymes from these two groups of organisms may well represent fundamentally different evolutionary products.

Not enough work has been performed on the chemical nature of β-lactamases from Mycobacteria to be able to say anything yet about their relationship to other β-lactamases.

In summary, therefore, β-lactamases are a widely distributed group of enzymes. Some of the enzymes in the group from gram-positive species have some structural features in common, and the same may be said for enzymes from gram-negative species, although there is less firm information for the latter. On the other hand, there seems to be little in common between enzymes from gram-positives and gram-negatives, except for the ability to hydrolyze the β-lactam bond and inhibition by iodine solutions. Within given species, the variants seem to be very similar molecules, with the differences amounting only to a few residues in the polypeptide sequences.

3.   β-LACTAMASE/ANTI-β-LACTAMASE INTERACTIONS

3.1   Methods of Investigation

The interactions of β-lactamases with their antisera have been studied in two main ways. On the one hand, many studies have involved neutralization of enzyme activity in solution (39,40). The other approach has been to use gel diffusion methods, all of which are basically variations on the original method of Ouchterlony. Few formal precipitation analyses in solution have been carried out, mainly because this technique requires relatively large amounts of antigen if the solubility product of the enzyme/antiserum precipitate is to be exceeded (41). Such amounts of sufficiently pure antigen are not easy to obtain.

3.1.1   Enzyme Neutralization Studies

In his early work on this subject, Cinader (39,42) concentrated his attention on enzyme/antiserum interactions that resulted in a diminution of the activity of the enzyme preparations—hence the term enzyme neutralization. However, he did recognize that some enzyme/antiserum complexes might be unaltered in activity. Furthermore, even where there was some degree of neutralization, the enzyme/antibody complex nearly always retained some "residual activity." Work with β-lactamases first showed that specific antisera could also enhance the activity of enzymes (27,43). For convenience, such stimulatory systems will be considered in this section, because these sera can be used in much the same way as those that neutralize.

The effect of antiserum on enzyme activity can give much detailed quantitative information about an enzyme. Moreover, the technique has the advantage that it can be used with impure enzyme preparations. In fact, usually the only requirement to be satisfied in making enzyme for this type of study is that it must be obtained in soluble form. With β-lactamases, this stipulation requires the use of extracellular enzyme from gram-positive species, or, by lowering the external osmotic pressure to shock out the enzyme, that of gram-negative species can be used.

Figure 1 shows a typical neutralization curve obtained when increasing quantities of a neutralizing serum are added to a given amount of enzyme; this curve is the so-called constant antigen titration. Several aspects of this curve are important for characterizing the enzyme. After a small "plateau zone," there is a linear fall in enzyme activity with increasing amounts of antiserum. This phase reflects the interaction of the antibody molecules with the enzyme and gives the "neutralizing titer" of the serum. The linear neutralization phase ends at the point, the equivalence point, where further additions of serum give no more neutralization. The amount of activity that remains at this point is the "residual activity" and is a measure of the activity of the enzyme-substrate complex. In practice, the neutralization slope and the residual activity are the parameters that are most valuable for quantifying the interaction of an enzyme and its specific antiserum.

Figure 2 shows a typical stimulation by antiserum. Again, several useful phases can be recognized. The slope of the curve, the stimulatory titer of the antiserum, is analogous to the neutralizing titer of inhibitory sera, and the equivalence point is the point at which maximum stimulation is obtained. The ratio of the activity when fully stimulated to that in the absence of serum is useful for quantitative studies and is known as the "stimulatory ratio."

It would not be appropriate here to indulge in a lengthy discussion of how enzyme/antiserum interactions may be used for quantitative studies on enzymes. However, three simple examples may give some idea of the possibilities. The residual activity may be used to calculate the amount of a β-lactamase that reacts with antiserum when it is present in a mixture with an enzyme that does not react. Figure 3 shows such an

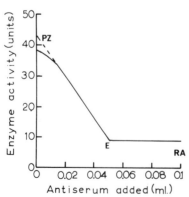

Fig. 1. Typical enzyme neutralization curve. PZ, plateau zone; E, equivalence point; RA, residual activity.

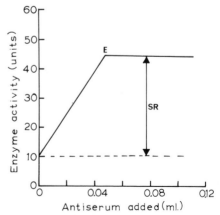

Fig. 2. Typical enzyme stimulation curve. E, equivalence point; SR, stimulatory ratio.

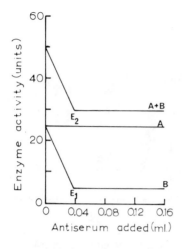

Fig. 3. The use of residual activity to calculate the proportion of penicillinase activity due to an interacting enzyme. Curve A, effect of antiserum on enzyme A (i.e., no reaction); curve B, effect of serum on enzyme B. Note that the equivalence point, E_1, defines the titer of the serum and the proportion of the activity that remains unneutralized. Curve C, effect of serum on a mixture of enzymes A and B. The characteristics of the serum, taken in conjunction with the equivalence point, E_2, allow the proportion of type-B enzyme of the mixture to be calculated.

example. If the characteristics of the interaction of the enzyme and the serum are known in detail, the relative proportions of the two enzymes may be calculated.

A second example is to test whether two enzyme preparations may be immunologically identical. If the enzymes are the same, a mixture of equal amounts of the two will give a neutralization curve indistinguishable from that obtained with twice the amount of reference enzyme. In particular, the neutralization titer (or stimulatory titer) should be unchanged, and the residual activity should be the same proportion of the initial activity as with a pure preparation of reference enzyme (Fig. 4).

The third example is the use of serum to detect the extent to which a mutant enzyme is a "mutein" (mutant protein; see Ref. 44) that has an abnormally low specific enzyme activity. Figure 5 shows such an example. If the residual activity is the same proportion as found with the wild-type enzyme, but the neutralization titer is, for instance, one fifth of normal, the likely conclusion is that the mutein is a molecule with one fifth the specific enzyme activity of the wild-type enzyme. Other examples of these techniques will be encountered later in this chapter.

The reaction between an enzyme and a specific antibody is not always simple. In certain cases, additions of small amounts of sera stimulate activity, whereas higher antibody concentrations neutralize. One example of this type of biphasic response is given by Pollock, who discusses the interaction of *B. cereus* 569/H type-I penicillinase with specific antiserum when methicillin is used as the substrate (see Fig. 4 and Ref. 43). Such titration curves are difficult to use for quantitative studies but do stress how antiserum preparations may have complex effects on enzymes.

Another point to emerge from this particular experiment, and also from others of a similar kind, is that the response of an enzyme to antiserum may depend greatly on which substrate is present. In the *B. cereus* type-I penicillinase/antipenicillinase example already mentioned, the antiserum gives an uncomplicated neutralization curve with benzylpenicillin as the substrate but a biphasic response with methicillin or oxacillin (43). Examination of the effect of anti-*B. licheniformis* antiserum on type-749 enzyme reveals five distinct responses, depending on whether benzylpenicillin, 6-aminopenicillanic acid, methicillin, cephalosporin C, or benzylcephalosporin C is used as the substrate (43). In general, such responses are interpreted as being a consequence of the molecular flexibility of the enzyme, a trait particularly evident with β-lactamases. All

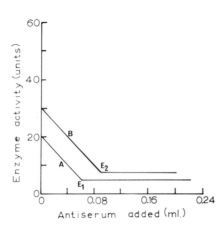

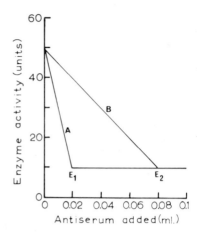

Fig. 4. The use of a neutralizing serum to compare two enzymes. Curve A, neutralization of pure enzyme A. This curve allows the titer of the serum and the position of the equivalence point, E_1, to be defined. Curve B, neutralization of pure enzyme A mixed with unknown enzyme activity. The titer, and thus the equivalence point, E_2, show the same relationship as with pure A.

Fig. 5. The use of antiserum interaction to characterize a mutant enzyme (or "mutein"). Curve A, neutralization of normal enzyme; curve B, neutralization of the mutein. Note that although the residual activity of the mutein is the same as that found with the wild-type enzyme, both the neutralization titers and the equivalence point (E_2) of the mutein indicate that this mutant enzyme is likely to have one third the specific activity of the wild type.

enzymes respond to the presence of substrates by a change in conformation, the so-called induced-fit hypothesis, and the operation of this system may clearly be effected differently with different substrates. In practice, only the relatively straightforward responses are easy to interpret in quantitative terms.

Before leaving this discussion of the use of neutralizing/stimulating antisera for enzyme studies, it is important to stress that various batches of antiserum may have very different properties, even when the same antigen preparation is used to raise them and the same immunization course is followed. An example has been described by Pollock (12). Two preparations raised against 569/H type-I penicillinase gave very different neutralizing characteristics, even though the same batch of enzyme was used as antigen in two different rabbits. An antigen preparation may even give rise to neutralizing sera on some occasions and stimulating ones on others (43a).

3.1.2   Gel-Precipitation Techniques

The gel precipitation techniques normally used to detect similarity between antigens have been applied to β-lactamases, and precipitation bands have even been obtained by growing bacterial cultures on agar surrounding a central well that contains antiserum. In this way, Pollock and associates (44) showed that some mutants of *B. licheniformis* characterized by abnormal expression of β-lactamase synthesized approximately normal amounts of an inactive, or sometimes partially active, protein, whereas the altered enzyme expression of other mutants was due to the production of lower than normal amounts of enzyme protein. In this way, therefore, they were able to distinguish structural from regulatory gene mutations (see Fig. 1 in Ref. 44). Further useful information may be obtained if the agar preparations are stained for β-lactamase activity after development of the precipitation bands. Under these circumstances, it is sometimes possible to distinguish the presence of an enzymatically inactive precipitate from ones that have substantial residual activity.

In general, these techniques are useful because they are easy to carry out. They have the disadvantage, however, that they do not give strictly quantitative information. It is not even possible, for example, to be sure how many amino acid changes are needed for a "reaction of identity" to be replaced by a "crossing" or "spurring" precipitate. In the experiments of Pollock and associates (44), single amino acid changes did not disrupt the reaction of identify (4), nor did the four or five amino acid differences that distinguish the β-lactamases from *B. licheniformis* strains 6346 and 749 have a major effect (4). At the other end of the scale, the cross reaction between *B. licheniformis* strain-749 anti-penicillinase and the enzyme from *B. cereus* 569/H, between which approximately 80% of the residues are equivalent (12a), is not one of identity, because some "spurring" occurs (44a). Perhaps the greatest difficulty in the use of antisera to detect the degree of similarity between two cross-reacting enzymes is the variability of antiserum preparations. Any rules that may emerge for the interactions of one serum preparation may not hold for another, and what holds for one enzyme/antiserum pair need not have wider relevance.

Some β-lactamases from gram-negative species have been examined by immunoelectrophoresis (45). This approach has the advantage, when used in conjunction with techniques for identifying which bands have enzyme activity, of allowing impure preparations to be examined for interaction with sera. The technique has the disadvantage, however, of requiring relatively large amounts of sera without giving quantitative information about the enzyme.

3.1.3   Other Techniques

Precipitation of antigens from solution has been used in one series of experiments to determine whether induced penicillinase in *B. cereus* was synthesized *de novo* after induction or whether conversion of an inactive "protoenzyme" to an active form occurred (41). The experiments involved the addition of nonradioactive carrier penicillinase to the radioactive enzyme synthesized after addition of inducer, followed by precipitation of all of the enzyme with excess antiserum (41). The experiments showed beyond doubt that penicillinase induction in *B. cereus* does involve *de novo* protein synthesis; however, this approach is of limited general use, because large quantities of pure nonradioactive antigen and of antiserum are needed.

3.2   Specific Enzyme/Antibody Interactions

3.2.1   *Bacillus cereus*

The first antisera to *B. cereus* penicillinase were prepared by Housewright and Henry (46), but the systematic immunologic analysis of this enzyme has been carried out by Pollock (12,43). The original experiments, in which anti-569/H type-I serum was used to compare enzyme from strains 569/H and 5/B, showed very little difference in neutralization characteristics (12). Similarly, anti-5/B type-I serum was unable to distinguish clearly between these two β-lactamases. However, if anti-569 serum is absorbed with purified 569/H type-I enzyme, and the resultant serum, after removal of the precipitate, is used to compare the 5/B and 569 enzymes, clear differences emerge (12). Similarly, absorption of anti-5/B serum with 5/B enzyme, followed by comparison of the 5/B and 569 enzymes with the absorbed serum also gives very different neutralization curves. Thus, the less avid antibodies that remain after absorption with homologous antigens give a lower titered serum with greater resolving power for small differences in the antigen.

These experiments were all carried out before the presence of type-II β-lactamase in culture supernatants of *B. cereus* was suspected (13). It is unlikely, however, that the small amounts of type-II enzyme in the preparations studied by Pollock will have much of an effect on the two important conclusions that may be drawn from these experiments: first, 569 type-I and 5/B type-I enzymes can be distinguished by immunologic means; second, absorbed sera have greater resolving power than do unabsorbed sera. Unfortunately, the exact molecular relationships between 569 type-I and 5/B type-I enzymes are still not clear, so it is difficult to interpret these experiments in terms of the amino acid differences between the two antigens.

Citri and his wife have published extensive studies on the conformational changes that occur in *B. cereus* β-lactamases in the presence of substrates (7,47,48). For example, the enzyme becomes more susceptible to inactivation by a variety of agents, such as heat, urea, and proteolytic enzymes, when in the presence of methicillin or oxacillin. Furthermore, the pH activity curves of this enzyme vary considerably from substrate to substrate, another indication of conformational changes (47). Combination with homologous sera is found to protect the enzyme against several of these effects. For example, with benzylpenicillin as the substrate, the enzyme is protected against inactivation by heat and by extremes of pH (47), while with methicillin and oxacillin, the enzyme-antiserum complex is more stable to urea and to proteolytic enzyme treatment, as well as to heating in solution (48). Because, with benzylpenicillin as the substrate, combination with the antiserum inhibits the enzyme, whereas in the presence of methicillin or oxacillin the activity of the enzyme is enhanced (43), complexing with antibody molecules can influence the performance of the enzyme either by increasing or decreasing its relative activity. Citri claims that both the variation of enzyme activity in the presence of substrates and that which is found on combination with antisera reflect the extreme conformational flexibility of this enzyme (7), and it is difficult to quarrel with this conclusion. It is, on the other hand, a little difficult to interpret these results unambiguously, because the necessary physicochemical studies needed to interpret these phenomena at a molecular level are not yet complete.

As mentioned earlier, a proportion of the β-lactamase synthesized by *B. cereus* strains is cell bound and is more sensitive than the exoenzyme to inhibition by iodine. The immunologic reactions of this material, the so-called γ-penicillinase, have been investigated by Pollock (17). True γ-penicillinase does not react with anti-type-I 569 antiserum (17), and serum raised against purified γ-enzyme will not neutralize the α form, although it will precipitate the γ version (40).

Analogous studies with α enzyme that has been put through a so-called α to γ conversion by treatment with urea or guanidinium hydrochloride give similar results (40,48). This conversion leads to a loss of cross reaction with antiserum prepared against purified α enzyme. The extent to which this "γ-type" β-lactamase made from α enzyme by chemical means is the same as "true" γ-penicillinase obtained from *B.*

cereus cells has been the subject of much debate (see, for example, Refs. 39,47,48). Nevertheless, these results do suggest strongly that the conformation of an antigen is more important than its primary sequence in determining its reaction with antiserum. Of course, these two factors are not always unconnected; for the α to γ conversions reported by Citri, however, the mildness of the chemical treatment involved makes it extremely unlikely that any change in primary sequence occurs. It is worth mentioning here that the α to γ conversion also increases the activity of the enzyme against one of its substrates (methicillin in this case) about 10-fold, but this conversion has no significant effect on the rate of benzylpenicillin hydrolysis (48).

The effect of amino acid analog incorporation on enzyme/antiserum interactions has been studied with *B. cereus* 569/H penicillinase (49,50). The analog used was *p*-fluorophenylalanine (a phenylalanine analog) (51), and the kinetics of incorporation were adjusted so that the majority of the phenylalanine residues of the enzyme were substituted with the fluoro analog. The neutralization of the "analog enzyme" with anti-569 type-I serum was somewhat impaired when compared with normal anti-569 type-I serum, and a much enhanced plateau zone (see Fig. 2) was observed. The latter observation suggests that amino acid analog incorporation produces a protein in which higher concentrations of antibodies are needed to initiate the enzyme/antiserum interaction.

3.2.2   *Bacillus licheniformis*

The original immunologic study of *B. licheniformis* penicillinases was carried out by Kushner (52), even though he thought he was studying isolates of *B. subtilis*. The interaction of anti-749 serum with penicillinase from strain 749/C produced a typical neutralization curve of the type shown in Fig. 2 when benzylpenicillin was used as the substrate. Antiserum to the enzyme from strain 6346, on the other hand, barely neutralized the homologous enzyme (52). With the 749 penicillinase/antipenicillinase system, the response obtained was very substantially effected by the choice of substrate used to assay the enzyme (43). For example, with benzylpenicillin, cephalosporin C, and benzylcephalosporin C, the serum neutralized in a straightforward manner, but with 6-aminopenicillanic acid, the response was biphasic, as manifested by a phase of stimulation followed by one of neutralization (see Fig. 6). With methicillin, the response was purely stimulatory.

The use of anti-749 and anti-6346 sera against the heterologous enzymes produced some effects on activity, but the results were somewhat bizarre (43). First, anti-6346 serum proved to be much more effective at neutralizing the 749 enzyme than the homologous 6346 enzyme. In practice, the neutralization titer of this enzyme against the enzyme from strain 749/C was about four times greater than when tested against the 6346 enzyme. When anti-749 serum was used against the enzyme from strain 6346, a range of responses was obtained, again depending on the substrate used. Whereas the interaction between 749/C enzyme and 749 antiserum produced uncomplicated neutralization (see above), the 6346/anti-749 interaction was biphasic, as was the response when 6-aminopenicillanic acid was used as the substrate. Methicillin gave a purely stimulatory response, whereas with the two cephalosporin substrates, a standard neutralization was obtained. Thus, antisera raised against these two types of enzyme from *B. licheniformis* clearly show some cross reaction when tested against homologous enzymes, but it is difficult to quantitate the extent of this interaction (43). Studies on the

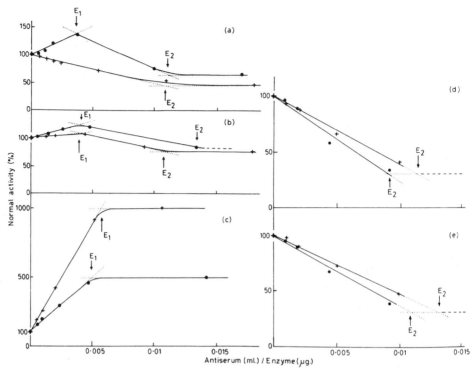

Fig. 6. Comparative antibody/enzymatic activity titration curves for penicillinase from *B. licheniformis* strains 749 (+) and 6346 (●) with antistrain 749 penicillinase, by use of five different substrates. (a), Benzylpenicillin; (b), 6-aminopenicillanic acid; (c), methicillin; (d), cephalosporin C; (e), benzylcephalosporin C. The equivalence points for the activating and inhibiting effects of the antiserum (calculated as described in the text) are indicated by E_1 and E_2, respectively. [From Pollock (43). By permission of the publishers of *Immunology*.].

structure of the enzymes from strains 749/C and 6346/C have shown that the two types of enzyme differ only in four or five amino acids in their primary sequence (4), although it would be difficult to believe that this is the case from the variety of reaction patterns obtained with antisera. In fact, these studies merely reinforce the view that the conformational state of the enzyme is much more important in determining antigen/antibody interactions than is primary sequence (see later).

Antisera to *B. licheniformis* penicillinases have also been used to examine the effect of changes in primary sequence on enzyme/antiserum interactions. After having excluded the mutants that were probably regulatory (because loss of enzyme activity was associated with loss of cross-reacting protein), four mutants of strain 749/C and three of strain 6346/C were selected for further analysis. The results are summarized in three Figures (Figs. 3a, 3b, and 3c) in the paper by Pollock and associates (44). Except for one mutant, addition of serum to the mutant enzymes minimized the effect of the mutational change on the activity of the enzyme; this phenomenon extended to measurements with three substrates. This unexpectedly uniform result can only mean that the effect on enzyme activity of the amino acid substitution in the primary sequence is due to the consequent change in conformation rather than to any altered chemical properties caused by the amino acid switching. Were this not so, it is difficult to see how complexing with antiserum could give a generalized enhancement of activity. Probably what happens is that the changed primary sequence leads to a more flexible enzyme molecule, and this increased flexibility, in turn, leads to lowered ability to hydrolyze the substrate. Interaction with antiserum may, however, limit the flexibility of the enzyme molecule, so that hydrolytic action against the substrate is more effective.

As mentioned previously, about 50% of the penicillinase activity is cell bound in exponentially growing cultures of B. *licheniformis* 749/C (32). This enzyme reacts with anti-749 serum, but the titers shown are reduced when compared with the response of the exoenzyme (53). Some of the cell-bound enzyme in B. *licheniformis* is attached to membrane vesicles (21,22), and, while in this location, the enzyme seems to be attached to a phospholipid component (54). Early studies had shown that the cell-bound enzyme in this strain could be liberated from the cells by treatment with trypsin (52) and that the liberated enzyme seemed to have properties, including size, that were very similar to those of the natural exoenzyme (20), although it now seems clear that the cell-bound material is not an obligate intermediate in the flow of the extracellular form from its site of synthesis. Lampen and colleagues have now shown that trypsin will liberate the β-lactamase from its complex with phospholipid (54), and this phenomenon may well be responsible for the trypsin-liberated material studied earlier (20). Trypsin-liberated enzyme has the same size and amino acid composition as the true exoenzyme, except that the N-terminal lysine is missing (20). Genetic experiments suggest, however, that the cell-bound enzyme before trypsin liberation may have several additional residues at its amino terminus, at least in the form in which it is synthesized on the ribosome (55). Few immunologic examinations of these various forms of B. *licheniformis* β-lactamase have yet been carried out. All that seems certain is that the trypsin-liberated material has a reaction with anti-749 serum that is indistinguishable from that of the true exoenzyme (4).

3.2.3 *Staphylococcus aureus*

Antiserum raised against purified extracellular β-lactamase from S. *aureus* (27) carrying an pI penicillinase plasmid has been used to investigate enzyme/antiserum reactions among penicillinases from other staphylococcal strains (25). Originally, three types of response were found, and recently a fourth type has been added (26). These four responses have been used to distinguish four types of staphylococcal β-lactamase: types A, B, C, and D (25,26).

Figure 7 shows typical reactions of these four enzyme types. With types A, B, and D, antiserum prepared against type A (the product of pI-type penicillinase plasmids; Ref. 56) gave a stimulatory response, in which the total activity of the enzyme/antiserum complex was about 4.5-fold as great as that of the untreated enzyme (25). Type-D enzyme gave a similarly general response, with a stimulation titer about half that found with type-A antiserum acting on type-A β-lactamase and a maximum stimulation of about 1.5-fold (26). Type-B enzyme gives a stimulation titer of about 5% of that found with the type-A enzyme and homologous serum, but addition of sufficient serum can produce the full increase of 4.5-fold (25). Type-C enzyme does not seem to react with anti-type-A serum when tested for its effects on enzyme activity (25). So far, only anti-type-A sera have been prepared, so homologous cross reactions with other enzyme types have not been tested.

Purification of the type-B enzyme, mostly synthesized by phase group-II strains of S.

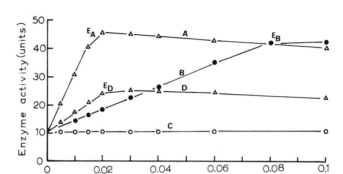

Fig. 7. The interaction of anti-A-type staphylococcal penicillinase serum with the various types of staphylococcal β-lactamase. A, B, C, and D are the four types of staphylococcal β-lactamase; E_A, E_B, and E_D are the equivalence points for the A, B, and D types of the enzyme.

aureus (57), shows that the enzyme has a specific activity of about 20% of that of the type-A enzyme, and consequently the interaction with anti-type-A serum can be interpreted fully. If the stimulation curve is replotted, not in terms of enzyme *activity* but in terms of enzyme *protein* (assuming a specific enzyme activity of about 1.5 units/ μg of protein; Refs. 25,27), the curves for the interaction with type-A and -B enzymes are seen to be superimposable (Fig. 8). Therefore, the interaction of anti-type-A serum with type-B enzyme protein is indistinguishable from the homologous interaction; the difference in stimulation characteristics are entirely due to the difference in specific enzyme activity.

So far, it is not possible to perform similar experiments with the type-D enzyme, because it has not been purified adequately. Nevertheless, the fact that this enzyme is only stimulated 1.5-fold argues that the difference from the homologous reaction must be more than just a matter of the specific activity of the enzyme.

Type-C enzyme does not seem to interact with anti-type-A serum, as assessed by the effect on enzyme activity. However, precipitation analysis shows that anti-type-A serum does react with type-C enzyme; the precipitation titer of the serum is about one third of the value obtained with type-A enzyme (43a).

These stimulatory responses of staphylococcal penicillinase are normally interpreted as being due to a conformational change induced by combination with antiserum or to locking of the enzyme in one of several potentially interconvertible conformations. Certainly, this enzyme seems to be just as flexible as the β-lactamases of other gram-positive species.

Perhaps the most clear-cut example of this phenomenon concerns the effect of anti-serum on the inactivation of staphylococcal penicillinase by the "new" penicillins, notably methicillin (58). Type-A staphylococcal penicillinase has a very low affinity for methicillin (59), and, moreover, this substrate inhibits the enzyme in a manner that is not readily reversible (60). This inhibition may be blocked by the presence of benzylpenicillin, but as soon as the benzylpenicillin itself has been destroyed, the inhibition by methicillin asserts itself (58). Another method of protecting staphylococcal penicillinase from the inhibitory action of methicillin is to interact the enzyme with an excess of anti-type-A serum before adding the methicillin. Although the stimulation achieved with methicillin as the substrate is not as great as that produced with benzylpenicillin, the combination with the antiserum completely blocks the inhibitory effect of methicillin (60). So far, no detailed binding studies of methicillin with antiserum-protected enzyme have been published. Nevertheless, it seems likely that methicillin inhibits by achieving a lethal "induced fit" of enzyme to substrate and that an enzyme-antiserum complex prevents this disadvantageous change without destroying the ability of the enzyme to act.

Before leaving this class of enzymes, it is important to point out that not all staphylococcal antisera stimulate. One rabbit inoculated with the same antigen preparation as that which gave the stimulatory sera described here and elsewhere gave a response that neutralized type-A staphylococcal β-lactamase by 35% (43a).

3.3   Cross Reactions among β-Lactamases of Various Gram-positive Species

Although Kushner could originally show no cross reaction between antisera raised against *B. licheniformis* β-lactamase when tested with *B. cereus* enzyme (52), a mild degree of cross reaction does occur. To show this cross reaction, it is necessary to use methicillin as the substrate. Under these circumstances, anti-749 serum stimulates the activity of 569/H type-I β-lactamase activity, with about 2% of the response shown in a reaction between 569/H type-I β-lactamase and its homologous serum in the presence of methicillin (43). Anti-*B. cereus* 569/H type-I serum did not cross react with *B. licheniformis* 749C β-lactamase (43).

Except in an early claim to the contrary (61), no cross reaction occurs between anti-staphylococcal β-lactamase serum and either *B. cereus* or *B. licheniformis* enzyme, nor do the reciprocal tests show any cross reaction (43). Furthermore, Pollock (44a) has recently looked for interactions between antisera to type-I β-lactamase from *B. cereus* and purified type-II enzyme and has found none. The low degree of cross reaction between anti-749 serum and *B. cereus* 569/H type-I enzyme therefore remains the only interspecies cross reaction yet confirmed among the enzymes from gram-positive species.

What is known of the primary polypeptide sequences of these enzymes certainly agrees with this observation. Only about 40% of the primary sequence of the *B. licheniformis* enzyme is "equivalent" to that of *S. aureus* (20), probably a large difference from the immunologic point of view. However, about 85% of the primary sequence of the *B.*

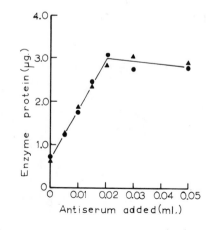

Fig. 8. A comparison of types -A and -B staphylococcal penicillinase when it was titrated with anti-type-A serum and the results were plotted in terms of enzyme protein rather than enzyme activity. Compare with Fig. 7.

licheniformis enzyme is "equivalent" to *B. cereus* type-I β-lactamases (12a), and this difference seems to be the basis for the mild degree of cross reaction found in the presence of methicillin. None of the "penicillinases" of the gram-positive species seem, on the basis of what is known at present, to have much structure in common with the "cephalosporinase" (type-II enzyme) from *B. cereus,* hence the absence of a cross reaction, even with anti-type-I serum.

No cross reaction between gram-positive enzymes and gram-negative sera, or vice versa, has yet been shown.

3.3.1   β-Lactamases of Gram-negative Species

Compared with the studies on gram-positive bacteria, relatively little has been published about the immunologic reactions of the enzymes from gram-negative species. Sera have been prepared against type-IIIa β-lactamase (35). With benzylpenicillin as the substrate, a good neutralization titer was obtained, but the residual activity of the enzyme-antibody complex was commonly as high as 35% of the initial activity of the enzyme. Similar sera have also been prepared against purified type-Ia enzyme from *Enterobacter cloacae* strain 214 (62). Again, neutralization was obtained with a relatively high residual activity of the enzyme-antibody complex.

Another approach to the preparation of specific antisera to β-lactamases from gram-negative species has been to raise sera against crude preparations of enzyme (usually made by disrupting enzyme-producing bacteria and then removing the debris by centrifugation) and then to absorb with a similar preparation made from a strain that has been mutated to remove β-lactamase synthesis (62a,72). In this way, antisera to types-Ia, -IIIa, and -IVc enzyme have been prepared. How adequate such preparations may be for neutralization analyses is not yet certain, because they have only been used to any extent for precipitation studies in gels or after isoelectric focusing in thin layers of polyacrylamide (63).

A combination of isoelectric focusing and antibody analysis has shown that certain of the β-lactamases from gram-negative species that were originally thought to represent a single type of enzyme are now known to contain minor variants. Thus, the "type-IIIa" enzyme, as specified by the R-factor R_{TEM} in *Escherichia coli,* and the "type-IIIa" enzyme, coded for by the plasmid RP1 in the same *E. coli* strain, show identical neutralization characteristics with anti-type-IIIa serum raised against purified type-IIIa enzyme (64). However, isoelectric focusing of these two versions of the enzyme shows that enzyme specified by R_{TEM} has an isoelectric point of 5.4, whereas that produced by RP1 focuses at pH 5.6. Two minor variants of type-IIIa enzyme therefore occur among bacteria that carry R factors (63,63a).

Similar studies on type-Ia enzyme reveal that this enzyme type also includes at least two minor variants. According to substrate profiles, the "type-Ia" enzymes synthesized by the *Enterobacter* strains 214 and P99 are indistinguishable (2,62), and both cross react identically with sera prepared against type-Ia enzyme from strain P99 (43a). Yet, the isoelectric point of the enzyme from strain P99 is pH 8.0, whereas the enzyme from strain 214 focuses at pH 8.2, and this difference persists when mixtures are run (63,63a).

The combination of isoelectric focusing techniques on thin-layer acrylamide gels with gel precipitation after focusing is complete is clearly an important tool in the comparison of the immunologic properties of various β-lactamases and is one that needs to be further exploited.

3.4   Cross Reactions among β-Lactamases of Various Gram-negative Species

In the original studies on the classification of β-lactamases from gram-negative species, the eight enzyme types considered were examined for cross reactions against sera obtained against purified type-IIIa enzyme (as specified by R_{TEM}) and anti-type-Ia serum (as specified by *E. cloacae* strain 214; Ref. 35). Type-Ia enzymes alone reacted with anti-type-Ia serum. Thus, the type-Ib cephalosporinases, on the basis of these experiments, do not cross react with sera to the type-Ia enzyme. Similarly, the only type of enzymes found to cross react significantly with anti-type-IIIa serum were those in group III, all of which were R-factor mediated (35). A slight degree of cross reaction was obtained with enzyme from *Enterobacter* strain 418 (35) and anti-type-IIIa serum, but this phenomenon has not been examined systematically any further.

Before any really valid statements can be made about the extent of immunologic relationship between the β-lactamases from gram-negative bacteria, sera must be raised against all of the enzyme types and then used to test the degree of cross reaction in all combinations, and preferably also with a range of substrates. Such studies will be no simple task. In the meantime, one can perhaps be reassured that what is already known of the immunologic cross reactions among this group of enzymes tends to confirm the various classifications based on substrate profile data.

No cross reactions between antisera raised to enzymes from gram-negative bacteria have yet been shown with enzymes from gram-positive species.

3.4.1   Mycobacteria

Little immunologic work on the β-lactamases of this group of organisms has yet been reported. Perhaps the only study of any significance is that of Kasik and colleagues (31), who studied the precipitation bands in agar obtained when crude enzyme preparations that contained β-lactamase activity from *Mycobacterium tuberculosis* R_1R_v and two strains of *Mycobacterium smegmatis* were tested against sera prepared against these crude preparations. Antisera absorbed with the enzyme preparations were then used to show which precipitation bands disappeared and could therefore be attributed, by negative means, to the penicillinase/antipenicillinase interaction. The only conclusions that can be drawn from this paper is that antisera to *M. tuberculosis* β-lactamase did not cross react significantly with the enzyme from *M. smegmatis*, and vice versa; quantitative neutralization studies are needed. These studies will not be easy to complete, with enzymes from a group of organisms so notoriously difficult to work with.

4.   PROTEIN STRUCTURE AND REACTION WITH SPECIFIC ANTISERA

Despite the relatively large amount of information available, it is still impossible to obtain a clear picture of the mechanism by which changes in the primary sequence of an enzyme affect its interaction with antiserum. Among β-lactamases, the most complete studies concern the interaction of *B. licheniformis* 749 enzyme with its specific antiserum (44, 52, 53, 65); yet, even here, it is difficult to draw any firm conclusions. It seems probable that the effect of primary sequence on antiserum interactions will only be completely understood when the effect of sequence on enzyme conformation is fully worked out.

The simplest variants of *B. licheniformis* 749 enzyme to be studied in detail with specific antisera are the four "muteins" (mutant proteins; Ref. 44) isolated by Dubnau and Pollock (66). All are probably due to single amino acid transpositions, although the exact nature and positions of the changes are not yet clear (4). In all four muteins, titration with antiserum gives unchanged equivalence points when compared with the wild-type enzyme, even when the effect of the serum on enzyme activity varies widely (44). Pollock takes the constancy of the equivalence points found in these titrations to imply that the interaction between the muteins and antiserum has caused no change stoichiometrically and therefore that the changes in primary sequence have not affected the nature of the binding sites of enzyme to antibody but that the conformational flexibility of the enzyme is much increased by the changes.

Certainly four, and may be five, differences in primary sequence distinguish 749 β-lactamase from the enzyme from *B. licheniformis* strain 6346 (67,68), but even these changes do not seem to be sufficient to alter the equivalence of enzyme and antiserum when 6346 β-lactamase is titrated with anti-749 serum (44).

At the other end of the scale, anti-749 serum shows no detectable interaction with purified *S. aureus* β-lactamase type A (44). About 40% of the primary sequence of the staphylococcal and *B. licheniformis* 749 lactamases are the same (20), but the 60% difference is apparently sufficient to completely disrupt recognition. An intermediate position in this series is occupied by *B. cereus* type-I β-lactamase. In the extensive regions of this enzyme that have been sequenced, about 80% of the residues occupy positions equivalent to those in *B. licheniformis* 749 β-lactamase (12a). In this case, a trace of cross reaction can be seen between *B. cereus* type-I enzyme and anti-749 serum but only when methicillin is used as the substrate (43).

One of the most interesting observations made with 749 enzyme/anti-749 serum interactions is that addition of antiserum to mutein β-lactamases seems to make the enzyme behave much more as though it were wild-type in properties (44). Because there is fairly good evidence that the antibody-binding sites on the mutein are unaffected by the amino acid transpositions studied so far (the equivalence of mutein to antiserum is unchanged; see above), the main factors that reduce the specific activity of 749 muteins are likely to be conformational changes rather than changes of residues in the enzyme-active center. This pattern would certainly fit well with the observation that addition of sera to many β-lactamases significantly alters substrate profiles under circumstances where no differences in primary sequence can exist.

All of this evidence suggests, therefore, that the enzyme from *B. licheniformis* 749, and probably all other β-lactamases, are extremely flexible molecules in which conformational changes are readily induced both by substrates and by other environmental changes. These changes cannot be irreversible, because sera often bind with unaffected equivalence. However, the changes of conformation that follow this interaction may themselves have profound effects on the resultant catalytic effect of the protein. The interactions we have been considering here are all manifestations of responses in a three-component system, enzyme-substrate-antibody, and it is not surprising that the effects produced can be complex and difficult to interpret, particularly when it is remembered that all of the antiserum preparations used contain a large number of antibody molecules, each with different properties.

5. PENICILLINASE/ANTIPENICILLINASE AS A MODEL SYSTEM

The detailed analysis of the factors that affect enzyme/antibody interactions has been greatly hindered by the lack of a suitable model system with which to work. It seems likely, however, that the *B. licheniformis* 749 β-lactamase/anti-β-lactamase system may now help to fill in the gaps in our knowledge of such interactions. An alternative might be the staphylococcal penicillinase/antipenicillinase reaction, but, in the end, genetic analysis with *B. licheniformis* is likely to be easier.

If one first considers the enzyme component of the reaction, several conditions must be satisfied if the system is to be adequate. First, the assay of the enzyme must be simple, and a wide range of substrates must be available. The assay of β-lactamases has been investigated by many workers, and reliable and simple methods that can be used in the presence of serum proteins are well worked out (69). These assay techniques have even been automated, so the number of individual measurements needed is unlikely to ever be a limiting factor in the design of an experiment (70). A large number of penicillins and cephalosporins are commercially available, and the β-lactamase of *B. licheniformis* 749 is active against most of them (see Table 3 and Refs. 1,4,6,32).

The second requirement of the enzyme in a model system is that any changes in its molecular nature should be capable of rapid detection and characterization. Because the complete sequence of the wild-type enzyme is known (2), single amino acid transpositions in *B. licheniformis* β-lactamase are relatively easy to work out and assign to their location in the primary sequence (4). Even multiple differences of the type that distinguish the 749 and 6346 enzymes can be elucidated without very great difficulty.

The third requirement is that one should be able to generate the necessary mutant enzymes readily and analyze the genetic changes involved. Pollock and colleagues have already shown that it is possible to isolate muteins from *B. licheniformis* β-lactamase very readily after several different mutagenic treatments (66), and the ability to transform this species of *Bacillus* gives one the ability to locate the genetic lesions by classic three-point recombination analysis (55,66,67). In fact, it is this last point that makes the enzyme from *B. licheniformis* 749 preferable to that from *S. aureus*. The *Bacillus* enzyme is specified by a chromosomal gene and is therefore easier to map than its staphylococcal counterpart. The staphylococcal β-lactamase gene is commonly carried on a plasmid of molecular weight less than 60×10^6 daltons (56), and the ordering of mutations on such structures raises all of the difficulties associated with mapping genes on small circular replicons (71).

When one considers the antiserum component of the system, the most important points are that it should have a high titer, and that interaction with the enzyme should give either a large stimulation or neutralization, with clear-cut equivalence points. *B. licheniformis* β-lactamase readily gives rise to sera in rabbits that require diluting at least 100-fold (52). An accurate stimulation or neutralization curve can then be obtained with less than 0.2 ml of dilute serum.

In summary, therefore, the generation of a suitable range of *B. licheniformis* β-lactamase muteins should allow the effects of antiserum combination with this enzyme to be classified on a strictly molecular basis. If the evidence so far obtained from experimental work are any indication, it can be said that only a restricted number of amino acid changes will directly alter the binding of antibody molecules with the enzyme, but the great majority of muteins will show some variation in substrate profile and in the effect of combination with antiserum. The interest will center on whether given types of alteration in neutralization or stimulation map together in particular areas of the gene, or whether the insertion of particular types of amino acids have the greatest effect on the interaction of this interesting enzyme with its antisera.

REFERENCES

1. N. Citri, in P. Boyer, Ed., *The Enzymes,* 3rd edit., Vol. 4, Academic Press, New York and London, 1972, p. 23.
2. M. H. Richmond and R. B. Sykes, in A. H. Rose and D. Tempest, Eds., *Advances in Microbial Physiology,* Vol. 9, Academic Press, London and New York, 1973, p. 31.
3. D. R. Thatcher, in J. Hash, Ed., *Methods in Enzymology,* Vol. 43, Academic Press, New York and London, 1975, p. 640.
4. D. R. Thatcher, in J. Hash, Ed., *Methods in Enzymology,* Vol. 43, Academic Press, New York and London, 1975, p. 653.
5. M. H. Richmond, in J. Hash, Ed., *Methods in Enzymology,* Vol. 43, Academic Press, New York and London, 1975, p. 664.
6. N. Citri and M. R. Pollock, in F. F. Nord, Ed., *Advances in Enzymology,* Vol. 28, Interscience, New York, 1966, p. 237.
7. N. Citri, in A. Meister, Ed., *Advances in Enzymology,* Vol. 37, Interscience, New York, 1973, p. 397.
8. M. R. Pollock, in G. E. W. Wolstenholme and C. M. O'Conner, Eds., *Drug Resistance in Micro-organisms,* J. and R. Churchill, London, 1957, p. 78.
9. M. H. Richmond, *Brit. Med. Bull.,* **21,** 260 (1965).
10. M. Kogut, M. R. Pollock, and E. J. Tridgell, *Biochem. J.,* **62,** 391 (1956).
11. M. R. Pollock, A. M. Torriani, and E. J. Tridgell, *Biochem. J.,* **62,** 39 (1956).
11a. R. P. Ambler, Personal communication (1974).
12. M. R. Pollock, *J. Gen. Microbiol.* **14,** 90 (1956).
12a. D. R. Thatcher, Personal communication (1900).
13. S. Kuwabara and E. P. Abraham, *Biochem. J.,* **103,** 27c (1967).
14. S. Kuwabara and E. P. Abraham, *Biochem. J.,* **115,** 859 (1969).
15. S. Kuwabara, *Biochem. J.,* **118,** 457 (1970).
16. M. R. Pollock and J. Fleming, *J. Gen. Microbiol.,* **59,** 303 (1969).
17. M. R. Pollock, *J. Gen. Microbiol.* **15,** 90 (1956).
18. N. Citri, N. Garber, and M. Sela, *J. Biol. Chem.,* **235,** 3434 (1960).
19. N. Citri and A. Kalkstein, *Arch. Biochem. Biophys.,* **121,** 720 (1967).
20. R. P. Ambler and J. Meadway, *Nature (London),* **222,** 24 (1969).
21. M. G. Sargeant, B. K. Ghosh, and J. O. Lampen, *J. Bacteriol.,* **96,** 1329 (1968).
22. B. K. Ghosh, M. G. Sargeant, and J. O. Lampen, *J. Bacteriol.,* **96,** 1314 (1968).
23. M. G. Sargeant and J. O. Lampen, *Proc. Nat. Acad. Sci. USA,* **65,** 962 (1970).
24. M. G. Sargeant and J. O. Lampen, *Arch. Biochem. Biophys.,* **136,** 167 (1970).
25. M. H. Richmond, *Biochem. J.,* **94,** 584 (1965).
26. V. T. Rosdahl, *J. Gen. Microbiol.,* **77,** 229 (1973).
27. M. H. Richmond, *Biochem. J.,* **88,** 542 (1963).
28. H. C. Neu, *Biochem. Biophys. Res. Commun.,* **32,** 258 (1968).
29. M. H. Richmond and N. A. C. Curtis, *Ann. N.Y. Acad. Sci.,* **235,** 553 (1974).

29a. Kintomichalon, P. M., E. G. Papachristrou, and E. M. Levis, *Antimicrob. Agents Chemothr.*, **6,** 60 (1974).

30. J. I. Kasik and L. Peacham, *Biochem. J.,* **107,** 675 (1968).

31. J. S. Thimpson, C. D. Severson, N. A. Stearns, and J. E. Kasik, *Infect. Immunity,* **5,** 542 (1972).

32. M. R. Pollock, *Biochem. J.,* **94,** 666 (1965).

33. M. R. Pollock, *Brit. Med. J.,* **IV,** 71 (1967).

34. N. Dalton and M. H. Richmond, *Biochem. J.,* **98,** 204 (1966).

35. G. W. Jack and M. H. Richmond, *FEBS Lett.,* **12,** 30 (1970).

36. M. H. Richmond, G. W. Jack, and R. B. Sykes, *Ann. N.Y. Acad. Sci.,* **182,** 243 (1972).

37. J. W. Dale, *Biochem. J.,* **123,** 493 (1971).

38. J. W. Dale and J. T. Smith, *Biochem. J.,* **123,** 500 (1971).

39. B. Cinader, *Ann. N.Y. Acad. Sci.,* **103,** 495 (1963).

40. M. R. Pollock, *Ann. N.Y. Acad. Sci.,* **103,** 989 (1963).

41. M. R. Pollock and M. Kramer, *Biochem. J.,* **70,** 665 (1958).

42. B. Cinader, *Annu. Rev. Microbiol.,* **11,** 371 (1957).

43. M. R. Pollock, *Immunology,* **7,** 707 (1964).

43a. M. H. Richmond, Unpublished observations (1974).

44. M. R. Pollock, J. Fleming, and S. Petrie, in *Antibodies to Biologically Active Molecules,* Proceedings of the 2nd Meeting of the Federation of European Biochemical Societies, Vienna, Vol. 1. Pergamon Press, Oxford and New York, 1966, p. 139.

44a. M. R. Pollock, Personal communication (1974).

45. G. W. Ross, in J. Hash, Ed., *Methods in Enzymology,* Vol. 40, Academic Press, New York and London, 1975, p. 678.

46. R. D. Housewright and R. J. Henry, *J. Bacteriol.,* **53,** 241 (1971).

47. N. Zyk and N. Citri, *Biochim. Biophys. Acta,* **159,** 317 (1968).

48. N. Zyk and N. Citri, *Biochim. Biophys. Acta,* **159,** 327 (1968).

49. M. H. Richmond, *Biochem. J.,* **77,** 121 (1960).

50. M. H. Richmond, *Biochem. J.,* **77,** 112 (1960).

51. M. H. Richmond, *Bacteriol. Rev.,* **26,** 398 (1962).

52. D. Kushner, *J. Gen. Microbiol.,* **23,** 381 (1960).

53. D. J. Kushner and M. R. Pollock, *J. Gen. Microbiol.,* **26,** 255 (1961).

54. T. Sawai, L. Crane, and J. O. Lampen, *Biochem. Biophys. Res. Commun.,* **53,** 523 (1973).

55. L. Kelly and W. J. Brammar, *J. Mol. Biol.,* **80,** 135 (1973).

56. M. H. Richmond, in A. H. Rose and J. F. Wilkinson, Eds., *Advances in Microbial Physiology,* Vol. 3, Academic Press, New York and London, 1968, p. 43.

57. M. H. Richmond, M. T. Parker, M. P. Jevons, and M. John, *Lancet,* **i,** 293 (1964).

58. A. Gourevitch, T. A. Pursiano, and J. Lein, *Nature (London),* **195,** 496 (1962).

59. R. P. Novick, *Biochem. J.,* **83,** 229 (1962).

60. K. G. H. Dyke, *Biochem. J.,* **103,** 641 (1967).

61. H. K. Rhodes, M. Goldner, and R. J. Wilson, *Can. J. Microbiol.,* **7,** 355 (1961).

62. T. D. Hennessey and M. H. Richmond, *J. Gen. Microbiol.,* **109,** 469 (1968).

62a. M. Matthew, A. M. Harris, M. Marshall and G. W. Ross, *J. Gen. Microbiol.,* **88,** 169 (1975).

63. G. W. Ross, in J. Hash, Ed., *Methods in Enzymology,* Vol. 43, Academic Press, New York and London, 1975, in press.

63a. G. W. Ross, Unpublished results (1975).

64. R. B. Sykes and M. H. Richmond, *Nature (London),* **226,** 952 (1970).

65. M. R. Pollock, *Ann. N.Y. Acad. Sci.,* **151,** 502 (1968).

66. D. Dubnau and M. R. Pollock, *J. Gen. Microbiol.,* **41,** 7 (1965).

67. D. Sheratt and J. F. Collins, *J. Gen. Microbiol.,* **76,** 217 (1973).

68. D. Sheratt, Ph.D. Thesis, University of Edinburgh, Edinburgh, Scotland.

69. G. W. Ross and C. H. O'Callaghan, in J. Hash, Ed., *Methods in Enzymology,* Vol. 43, Academic Press, New York and London, 1975, p. 69.

70. E. B. Lindstrom and K. Nordstrom, *Antimicrob. Agents Chemother.,* **1,** 100 (1972).

71. F. W. Stahl, *J. Cell. Physiol.,* **70** (Suppl. 1), (1967).

72. G. W. Ross and M. G. Boulton, *J. Bacteriol.,* **112,** 1435 (1972).

CHAPTER 3

A DEMONSTRATION OF EVOLUTIONARY RELATIONSHIPS AMONG THE LACTIC ACID BACTERIA BY AN IMMUNOCHEMICAL STUDY OF MALIC ENZYME AND FRUCTOSE DIPHOSPHATE ALDOLASE

JACK LONDON

1. INTRODUCTION

Antibodies are generally prepared against purified enzymes for one of two purposes. A specific immune serum can be used as a probe to characterize the biochemical and structural properties of the immunizing antigen and related sets of proteins or it can be used to measure structural similarities among homologous proteins to assess natural relationships between organisms that possess these isofunctional enzymes.

In connection with the former, specific immune sera have been used successfully to measure the dissociation and reassociation of the nonidentical subunits of aspartate transcarbamylase (1,2), to follow physical modification in the structure of hemoglobin A (3), to detect conformational changes in lactate dehydrogenase (4), to discriminate between specific substrate sites of alcohol dehydrogenase (5), and to measure amino acid modifications of genetically altered enzymes (6,7). However, it is in the area of serotaxonomy that the enzyme-antienzyme systems are being intensively exploited with the aim of enlarging our understanding of natural relationships among various organisms. Although the literature that deals with the serologic taxonomy of eukaryotes is not extensive, it is growing (8); however, only recently have the same immunologic techniques been used to systematize prokaryotes. A summary of the enzymes and the microorganisms surveyed is presented in Table 1.

The following discussion describes attempts by our laboratory to develop a natural classification for the homofermentative lactic acid bacteria by use of a malic enzyme (E.C.1.1.1.39) and a class-II fructose diphosphate aldolase (E.C.4.1.2.13) as evolutionary markers.

Table 1. A Summary of Enzyme:Antienzyme Systems used to Establish Immunologic Relationships Among Prokaryotes

Enzyme	Bacterial group surveyed	Reference
Alcohol dehydrogenase	Methane-oxidizing bacteria	Patel et al. (9)
Alkaline phosphatase	Enterobacteriaceae	Wilson and Kaplan (10), Cocks and Wilson (11), Steffen et al. (12)
Adenosine triphosphatase	Micrococcaceae	Whiteside et al. (13)
DNA polymerase	Enterobacteriaceae	Tafler et al. (14)
Fructose diphosphate aldolase	Lactobacillaceae	London and Kline (15)
Glutamine synthetase	Enterobacteriaceae Pseudomonadaceae Azotobacteraceae	Tronick et al. (16)
Lactate dehydrogenase	Lactobacillaceae	Gasser and Gasser (17)
Malic enzyme	Lactobacillaceae	London et al. (18, 19)
Muconate-lactonizing enzyme	Pseudomonadaceae	Stanier et al. (20)
Muconolactonizing enzyme	Pseudomonadaceae	Stanier et al. (20)
Phycobilins	Cyanobacteria	Bogorad (21), Berns (22), Glazer et al. (23)
Ribulose diphosphate carboxylase	Hydrogen bacteria, sulfur bacteria, photosynthetic bacteria	McFadden and Denend (24)
Tryptophan synthetase	Enterobacteriaceae	Creighton et al. (25), Murphy and Mills (26), Rocha et al. (27)

2. METHODOLOGIC CONSIDERATIONS

2.1 Preparation of Immune Sera

For taxonomic studies, it is imperative that the immunizing antigen be homogeneous. The production of a monospecific antiserum from a pure protein allows a rapid comparison of related proteins in crude extracts of microorganisms or tissues, because any resultant immunologic reaction will be due solely to the interaction of the specific antibodies and their antigen. If the antigen is contaminated with other proteins, the results of some immunologic procedures will be ambiguous and difficult to interpret; this is especially true of double-diffusion experiments. In the studies cited here, protein homogeneity was established by polyacrylamide gel electrophoresis and sodium dodecyl sulfate (SDS) polyacrylamide gel electrophoresis, following a routine purification of the malic enzymes (18,19,28), and aldolases (15,29) from lactic acid bacteria. Subsequent to the preparation of the antisera, the specificity of the antigen-antibody system was tested by immunoelectrophoresis of both the pure enzyme and the crude cell-free extract that served as the starting material for the purification of the antigen. The formation of a single sharp precipitin arc (Fig. 1) with both sources of antigen indicates that the immunizing antigen was pure within the limits of detection.

Antisera were prepared in 6-month-old male New Zealand white rabbits. Generally, four intradermal injections, each of which contained between 150 and 400 μg of antigen protein, 100 μg of methylated bovine serum albumin, and Freund's complete adjuvant, were administered at weekly intervals. One to two intravenous injections of a mixture of

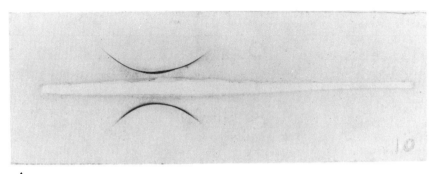

A.

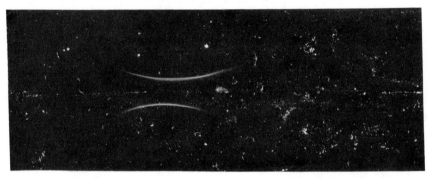

B.

Fig. 1. Immunoelectrophoresis of purified enzymes and their crude cell-free extract source. A: The upper well contains 8 μg of purified malic enzymes, while the lower well contains 30 μg of crude *Streptococcus faecalis* cell-free extract. Thirty micrograms of antiserum were placed in the center trough. The immune precipitate was stained with amino black. B: The upper and lower wells contain 12 μg of purified *S. faecalis* FDP aldolase and 40 μg of crude cell-free extract, respectively. The center trough contains 40 μl of a 1 : 5 dilution of antiserum.

the antigen and methylated bovine serum albumin over the following 2-week period would generally produce very-high-titer antisera.

2.2   Immunologic Techniques

The degree of immunologic relatedness between the homologous and heterologous antigens was estimated by the three following procedures: immunodiffusion, quantitative precipitation, and microcomplement fixation. Immunodiffusion or double-diffusion is generally performed according to the procedures of Ouchterlony (30) or Stollar and Levine (31). Because a cross-match between homologous and heterologous or heterologous and heterologous antigens can produce one of three basic precipitation patterns (15,17), the following discussion will serve to identify and define the three types of immune precipitates.

1.   A reaction of identical specificity occurs when a protein that possesses a complement of antigenic determinants identical to those borne by the homologous protein

interacts with the latter to produce confluent or fused lines of precipitation (Fig. 2a). A fused precipitate will also be produced if the cross-match pairs two heterologous antigens that share a smaller, but identical, complement of determinants with one another and the homologous protein. This is known as a reaction of apparent identity.

2.   A reaction of partial identity is characterized by a precipitate with a single spur (Fig. 2a). In this instance, the protein against which the spur is produced shares fewer common determinants with the reference (homologous) protein than does its neighbor. With appropriate cross-matches as controls, spurred precipitates can be used to establish a unidirectional line of evolution within a group of organisms.

3.   A reaction of nonidentity is characterized by double spurs (Fig. 2b), which indicate that the paired antigens share determinants with the homologous protein but very few or none with each other. Such a pattern indicates that the two proteins have diverged from one another relative to the reference antigen and are considered to be on separate and distinct lines of evolution.

Once a rough ordering of a set of related proteins has been determined by extensive immunodiffusion cross-matching, microcomplement fixation or quantitative precipitation is used to quantitate the antigenic differences within the set. Microcomplement fixation is usually the method of choice, because it requires far less antigen and antibody and the procedure is between 100 and 1000 times more sensitive than quantitative precipitation. Wasserman and Levine (32) described the microcomplement fixation procedure originally, and Champion et al. (33) have recently refined and updated the technique. The degree of relatedness between a homologous and heterologous protein was expressed as the index of dissimilarity (ID) by Sarich and Wilson (34). The index of dissimilarity is defined as the level to which the antibody concentration in the heterologous system must be raised so that the percentage of complement fixed by the heterologous antigen is equal to the percentage of complement fixed by the homologous protein. Minor modifications in the treatment of microcomplement fixation data have been suggested by Tafler et al. (14). Champion et al (33) have derived an equation that allows the comparison of microcomplement fixation results for those instances where it is not

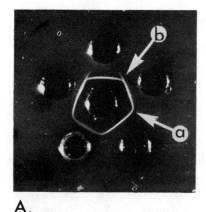

A.

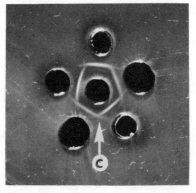

B.

Fig. 2. Immunodiffusion precipitation patterns. A: Confluent lines of precipitation reaction of identical specificity (a) and a single spurred precipitate or reaction of partial identity (b). B: Precipitate with crossed spurs or reaction of nonidentity (c).

possible to make the amount of complement fixed by the heterologous system equal to that fixed by the homologous system. The equation is as follows:

$$\log \text{ID} = \frac{Y_\text{H} - Y_\text{h}}{m} + \log \frac{X_\text{h}}{X_\text{H}}$$

where Y_H and Y_h are the percentages of complement fixed by the homologous and heterologous antigens, respectively, X_H and X_h are the antiserum concentrations used with the homologous and heterologous antigens, respectively, and m is the resultant slope of a semilogarithmic graph in which the log of the antiserum dilution used in the homologous antigen system is plotted against the percentage of complement fixed. The quantitative relationships can also be expressed as immunologic distance units (35), where the immunologic distance = log ID × 100. For the avian egg-white lysozymes, a change of approximately five immunologic distance units was equivalent to a substitution of 1% of the amino acids in the primary sequence of the enzyme. Although this relationship has not yet been confirmed with other proteins, the convention of expressing antigenic differences as immunologic distance units permits various sets of proteins to be compared with one another.

3. BIOCHEMICAL AND PHYSICAL PROPERTIES OF THE ANTIGENS

3.1   Malic Enzyme

The malic enzymes found in *Streptococcus faecalis* and *Lactobacillus casei* are inducible enzymes with estimated molecular weights of 65,000 (19,28). Both proteins appear to be single polypeptide chains by virtue of the fact that treatment with SDS does not dissociate them into smaller-molecular-weight components; however, the SDS treatment does alter the configuration of the molecule, as evidenced by a change in the molecular weight of the protein. In SDS polyacrylamide gels, the malic enzymes have an estimated molecular weight of 47,000. In addition to sharing several other biochemical characteristics, which include identical specificities for cofactor and metal ion activator, the isofunctional malic enzymes also respond to the same modifers of catalytic activity. Fructose-1,6-diphosphate (FDP), 6-phosphogluconate (6-PG), and 3-phosphoglycerate (3-PG) act as negative effectors for both enzymes; however, adenosine triphosphate only inhibits the streptococcal malic enzyme. A comparison of the biochemical properties of the enzymes is presented in Table 2.

   Protein charge differences that manifested themselves as distinctive electrophoretic mobility coefficients provided the only means of easily distinguishing the various malic enzymes. In anionic buffering systems (pH 8.8–9.1), all *S. faecalis* malic enzymes tested migrated with an R_f between 0.49 and 0.50. The *L. casei* malic enzymes were subdivided into two groups based on their net charge; the *L. casei* var. *casei* enzymes migrated with an R_f of 0.57–0.59, whereas the *L. casei* var. *rhamnosus* enzymes were far more electropositive and exhibited R_fs of 0.28–0.38 (19).

Table 2. Biochemical and Regulatory Properties of Malic Enzyme

Properties of the enzyme	Source of the enzyme	
	S. faecalis	L. casei
Estimated molecular weight	65,000	65,000
Cofactor specificity	NAD^+	NAD^+
K_m for substrate (mM)	0.11	0.42
K_m for cofactor (mM)	0.005	0.11
pH optimum	8.5	8.2–9.0
Metal requirement	Mn^{2+}	Mn^{2+}
Allosteric inhibitors		
FDP	$N = 3$[a]	$N = 1$
3-PG	$N = 1$	$N = 2$
6-PG	$N = 1.6$	$N = 2$
ATP	$N = 2$	No effect

[a] N = cooperativity coefficients determined by Hill plots.

3.2 Fructose Diphosphate Aldolase

The constitutively synthesized FDP aldolase found among the homofermentative members of the tribe Lactobacilleae vary considerably in size. Thus far, five molecular weight categories of this enzyme have been described (15), and one representative from the three most ubiquitous forms has been purified and characterized (29). The distribution of the five molecular weight categories of the lactic acid bacterial aldolases and other salient properties of the enzymes are shown in Table 3. Within the genus *Streptococcus*, the 56,000 molecular-weight aldolase is the most prevalent form, while the majority of lactobacilli possess the 118,000 molecular-weight enzyme. The pediococci, propionibacteria, and *Arachnia propionica* (see Table 3) possess the largest of the five forms of aldolase (mol wt = 176,000). Two forms of aldolase, the 76,000 and 135,000 molecular-weight enzymes, appear to be restricted to only a few species. The smaller enzyme is found in *Lactobacillus xylosus, Lactobacillus salivarius,* and *Lactobacillus plantarum,* and, as yet, only strains of *Streptococcus salivarius* have been shown to possess the 135,000 molecular-weight enzyme. Aldolases from several gram-positive nonspore-forming anaerobic bacteria that belong to the genera *Butyribacterium, Propionibacterium, Eubacterium,* and *Arachnia* are also distributed among the three major molecular weight groups.

The enzyme of *S. faecalis* ATCC 27792 was the first aldolase to be purified and used as a reference protein in the original lactic acid bacteria study (15); this 56,000 molecular-weight enzyme was readily dissociated with SDS into two subunits of identical size (mol wt = 28,000). Dividing the molecular weight of the four remaining size classes of aldolase by the molecular weight of the *S. faecalis* aldolase subunit yields values near the integers 3, 4, 5, and 6. It was subsequently learned (29) that the subunits of the *L. casei* and *Pediococcus cerevisiae* aldolases were slightly larger (mol wt = 29,500) than the streptococcal aldolase subunits. Nevertheless, immunologic simi-

Table 3. Distribution of the Five Molecular Weight Groups of Aldolase and Their Biochemical Characteristics

Molecular weight class	K_m for substrate[a] (mM)	Range of electrophoretic mobility coefficients	Distribution
56,000 (56,000–64,000)[b]	0.8	0.63–0.79	Major form in genera *Streptococcus* and *Eubacterium*; also found in *Lactobacillus curvatus, Microbacterium thermosphactum,* and *Pediococcus parvulus*
76,000 (70,000–80,000)	0.45	0.68–0.71	Found only in *Lactobacillus xylosus, Lactobacillus salivarius,* and *Lactobacillus plantarum*
118,000 (116,000–122,000)	0.45	0.32–0.50	Major form in genus *Lactobacillus*; also found in *Butyribacterium rettgeri*
135,000 (130,000–138,000)	0.8	0.75	Found only in strains of *Streptococcus salivarius*
176,000 (176,000–200,000)	0.38	0.50–0.55[c] 0.32–0.36[d]	Major form in genera *Pediococcus* and *Propionibacterium*; also found in *Arachnia propionica*

[a] K_m for FDP aldolase determined with enzymes from *Streptococcus faecalis, Lactobacillus xylosus, Lactobacillus casei, Streptococcus salivarius,* and *Pediococcus acidilacticii.*

[b] Range of molecular weights is given in parentheses.

[c] Range of R_fs for pediococci.

[d] Range of R_fs for propionibacteria and *Arachnia..*

larities among the five categories of aldolase were easily detected and measured. It appears, therefore, that whereas the quaternary structure of these enzymes has diversified markedly, the primary structure of the subunits has been conserved.

Based on their sensitivity to chelators such as ethylenediaminetetraacetate and *o*-phenanthroline, the immunologically related aldolases of the lactic acid bacteria appear to be typical prokaryotic class-II aldolases, as defined by Rutter (36). However, in addition to the antigenically cross-reactive class-II enzyme, strains of *L. casei* possess a serologically unrelated class-I enzyme whose function is as yet unknown (29). No allosteric effectors for the pure *S. faecalis* aldolase were found among a group of compounds that comprise the Embden-Meyerhof-Parnas and hexose monophosphate shunt pathway intermediate products, cofactors, and assorted mono-, di-, and triphosphate esters of nucleosides (15).

4.   ANTIBODIES AS STRUCTURAL PROBES

4.1   Detection of Conformational Changes in Malic Enzyme by Antimalic Enzyme Antibodies

As mentioned earlier, the catalytic activity of the streptococcal malic enzyme is regulated by three intermediate products of glucose catabolism: FDP, 6-PG, and 3-PG. The displacement of the substrate from the enzyme's active site was inferred from kinetic and heat inactivation studies (37). When specific antimalic enzyme serum was produced in rabbits, it became possible to confirm these results by an alternate method. The quantitative precipitin reaction described by Perrin (6) was modified to permit the interaction of the enzyme, its substrate, and inhibitory ligands prior to the addition of antibody. After the precipitate had formed, it was dispersed evenly, and the reaction mixture was tested for enzyme activity. The immune complex was then separated from the unreacted antibody and antigen by centrifugation, and the amount of free malic enzyme in the supernatant fluid was measured. The precipitate was washed several times with phosphate-buffered saline, and the protein content was determined. Although the interaction of enzyme and antibody produced an inhibition of catalytic activity, the addition of substrate (5 mM malate) enhanced the antibody-mediated inhibition of catalytic activity (Fig. 3). The effect of malate on the formation of the immune complex is also shown in Fig. 4. In the absence of malate, the zone of equivalence (as measured by the complete removal of malic enzyme activity from the reaction mixture after centrifugation and removal of the immune precipitate) was in the range of 0.3–0.8 units (Fig. 4). The equivalence zone was extended into the area that is normally the region of antibody excess by the addition of malate. However, the total amount of precipitable immune complex at equivalence is the same in the presence or absence of malate. It appears, therefore, that the interaction of enzyme with malate produces a conformational change

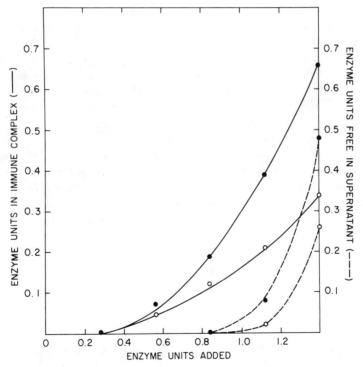

Fig. 3. The enhancement of antibody-mediated inhibition of malic enzyme by malate. The enzyme activity in the immune complex and precipitate-free supernatant are represented by the solid and dashed lines, respectively. The presence or absence of 5 mM malate is indicated by the open or closed circles, respectively.

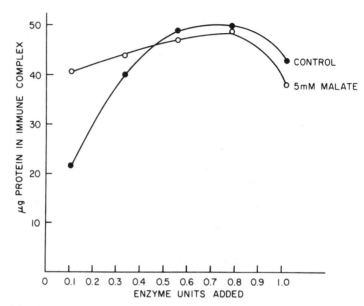

Fig. 4. The effect of malate on the protein content of the precipitated immune complex.

in the enzyme that increases immune complex formation with quantitatively less antigen than is needed in the absence of malate.

This antigen-antibody system provided an alternate method for probing the inhibitory ligand-enzyme interactions that were inferred from the kinetic experiments. Because FDP, 6-PG, and 3-PG appear to act only at the substrate-binding site of the malic enzyme (37), it seemed likely that each of the inhibitors would be able to reverse the enhanced antibody inhibition produced by malate. The addition of the respective inhibitors to the precipitin reaction mixture at concentrations identical to those used in kinetic experiments (2–20 mM) had no effect on the formation of the immune complex. However, after incubation of enzyme, antiserum, malate, and inhibitor, the malate-induced enhancement of antibody inhibition was partially reversed (Fig. 5). The levels of free enzyme that appeared in the supernatant fluid after removal of the precipitate by centrifugation were also displaced in the same order as those found for the whole reaction mixture.

Although microcomplement fixation appears to be the method of choice for detecting structural changes in proteins, the quantitative precipitin technique as employed in these studies also proved to be an effective tool for measuring ligand-induced conformational alterations in malic enzyme.

5. SEROTAXONOMY OF THE
LACTIC ACID BACTERIA

5.1 Natural Relationships Based on Immunochemical Studies of Malic Enzyme

Two physically and mechanistically distinct forms of malic enzyme are found among members of the tribe Lactobacilleae. The enzyme originally described by Ochoa and coworkers (38,39) appears to be relatively ubiquitous among lactic acid bacteria and, in this respect, contrasts sharply with the malic enzyme studied in our laboratory (18), which appears to be restricted to several species of this rather large tribe. The fact that the organisms that possess the latter enzyme also appear to possess the former suggests that the two enzymes do not perform the same physiologic function, especially because the growth conditions that induce one form repress the other (18).

Double-diffusion experiments with the use of anti-*S. faecalis* malic enzyme serum established that the malic enzymes from all strains of *S. faecalis* tested are immunologically identical; only confluent lines of precipitation were observed on Ouchterlony plates (Fig. 6a). In cross-matches between the *S. faecalis* and *L. casei* malic enzymes, single spurs were always produced against extracts of the latter (Fig. 6b). The antiserum failed to distinguish between the two electrophoretic mobility groups that corresponded to the subspecies *L. casei* var. *rhamnosus* and *L. casei* var. *casei* in immunodiffusion experiments. However, the two forms of malic enzyme were readily differentiated by microcomplement fixation (Table 4). This quantitative procedure established that the *L. casei* var. *rhamnosus* enzymes were less related antigenically to *S. faecalis* malic enzyme than were the *L. casei* var. *casei* enzymes.

Reciprocal experiments were performed with two malic enzymes from strains of *L. casei* var. *casei* that exhibit minor, but reproducible, differences in their electrophoretic mobility coefficients. Strain M-40 possesses a slightly more electronegative (basic) malic enzyme than does strain 64H. Ouchterlony double-diffusion experiments with anti-M40 and anti-64H malic enzyme sera easily differentiated the *L. casei* var. *casei* and *L. casei* var. *rhamnosus* malic enzymes (Fig. 6c), but neither serum detected any antigenic differences between the two minor *L. casei* var. *casei* subgroups. The order of relatedness as determined by immunodiffusion was *L. casei* var. *casei* > *L. casei* var. *rhamnosus* > *S. faecalis*. These findings agreed with and amplified the experiments described above. Microcomplement fixation experiments completely confirmed the immunodiffusion experiments (Table 4); because the indices of dissimilarity calculated from the quantita-

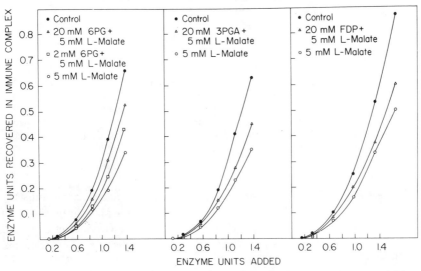

Fig. 5. Partial reversal of malate-enhanced antibody inhibition by allosteric inhibiters of malic enzyme.

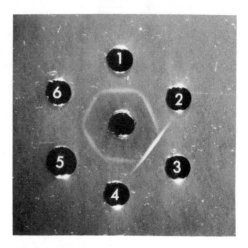

A.

B.

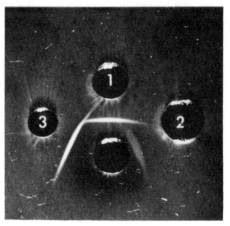

C.

Fig. 6. Immunodiffusion precipitin patterns produced by malic enzyme:antimalic enzyme interactions. A: Reactions of identical specificity with malic enzymes from five strains of *S. faecalis*. The homologous protein was placed in wells 1 and 4, and sufficient anti-*S. faecalis* malic enzyme serum was added to the center well to produce sharp lines of precipitation. B: Reactions of apparent identity among five strains of *Lactobacillus casei*; the homologous antigen is in well 3. The center well contains anti-*S. faecalis* malic enzyme. C: Reactions of partial identity between *L. casei* var. *casei* (well 3), *L. casei* var. *rhamnosus* (well 1), and *S. faecalis* (well 2) produced by a reaction with anti-*L. casei* var. *casei* malic enzyme placed in the center well.

tive data were essentially identical for both M-40 and 64H malic enzyme antisera, only one set of values is shown. The results are summarized by the phenogram shown in Fig. 7, which depicts the evolutionary relationships of the malic enzymes of *S. faecalis*, *L. casei* var. *casei* and *L. casei* var. *rhamnosus* and, by inference, the organisms that possess these enzymes.

Several tentative conclusions may be drawn from this study. It appears that not all biochemical and physical characteristics of enzymes are accurate indicators of structural

Table 4. Immunologic Distances of Lactic Acid Bacteria Malic Enzymes

Organism	Anti-*S. faecalis* malic enzyme		Anti-*L. casei* 64H malic enzyme	
	Index of dissimilarity	Immunologic distance	Index of dissimilarity	Immunologic distance
S. faecalis				
Strain: MR	1.00	0	20	130
H.3.1.	1.09	3.7	—[a]	
N83	1.05	2.1	—	
N37	1.00	0		
H69D5	1.00	0	—	
IA5	1.16	6.4	—	
H.9.1.	1.05	2.1	—	
L. casei var. *casei*				
Strain: 64H	52	171	1.00	0
M40	51	171	1.00	0
GC314	—[a]		1.18	7.1
GC316	—		1.20	7.9
L. casei var. *alactosus*				
Strain: OC17	50	169	—	
OC45	44	164	1.00	0
CL-16	–		1.00	0
L. casei var. *rhamnosus*				
Strain: OC91	78	189	5.70	75
CL-15	78	189	5.50	75
CL-11	72	185	11.0	104

[a] Not done.

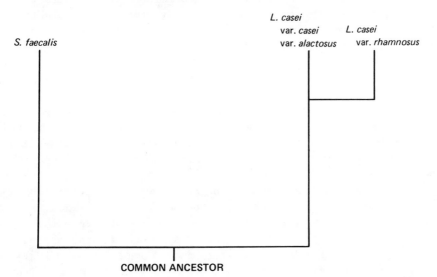

Fig. 7. A malic enzyme phenogram that shows the interrelationships among *S. faecalis* and the two subspecies of *L. casei*.

relatedness. For example, sets of proteins that exhibit similar electrophoretic mobility coefficients are generally considered to be closely related (40–43); however, for the malic enzymes, the electrophoretic mobility rates of the three antigenically related forms are relatively disparate. Similarities in the processes of enzyme induction (44) and regulation (45–47) are probably more accurate reflections of a common evolutionary history. Certainly, this was the case with the malic enzymes. Also, the malic enzyme findings begin to cast some doubt on the primacy of morphologic traits as taxonomic determinants. The structural homologies observed among the three electrophoretically distinct forms of malic enzyme suggest that the rod-shaped bacterium *L. casei* is phylogenetically related to the coccus-shaped *S. faecalis*. As will be seen shortly, it is no coincidence that this particular form of malic enzyme was found only in these two species of lactic acid bacteria. In the following section, the nature and extent of the relationship between these two morphologically distinct forms are amplified by a comparative study with FDP aldolase as yet another protein marker.

5.2   Natural Relationships among Homofermentative Lactic Acid Bacteria and Gram-positive Nonspore-forming Anaerobic Bacteria Based on Immunochemical Studies of Fructose Diphosphate Aldolase

Because all homofermentative lactic acid bacteria possess a FDP aldolase, the selection of this key Embden-Meyerhof-Parnas pathway enzyme as an evolutionary marker provided an opportunity to look for interspecies and intergeneric relationships within this large group of microorganisms. It was obvious from the outset that unless the primary structure of the aldolase had been conserved during the evolution of this group, there was little chance of success in any attempt to relate the various taxa that comprise the lactic acid bacteria with this enzyme. The FDP aldolase of *S. faecalis* (15) and *P. cerevisiae* (29) were purified to homogeneity, as judged by polyacrylamide gel electrophoresis, and used as antigens to prepare specific antisera in rabbits. The two groups of antisera were used to survey the aldolases of more than 50 species of bacteria that, on the basis of previous conventional taxonomic studies, were thought to be related. Because the titer of the anti-*S. faecalis* aldolase serum was approximately threefold greater than that of the anti-*P. cerevisiae* aldolase serum (1:90,000 versus 1:35,000 in microcomplement fixation assays), the latter was used essentially to confirm the extensive testing performed with the former antiserum.

Extensive pairwise cross-matches on Ouchterlony double-diffusion plates demonstrated the existence of antigenic homology between the *S. faecalis* aldolase and the aldolases of all homofermentative lactic acid bacteria tested; these bacteria included members of the genera *Streptococcus, Lactobacillus,* and *Pediococcus*. Immunologic relatedness was also observed between the reference aldolase and aldolases of certain members of the genera *Microbacterium, Eubacterium, Butyribacterium, Propionibacterium,* and *Acholeplasma*.

Immunodiffusion experiments with strains that represented 21 species of *Streptococcus* and included several unspeciated streptococci produced a classification scheme in which the streptococcal aldolases could be relegated to one of seven antigenic groups. Extensive cross-matching resulted in the appearance of only two types of precipitation patterns with the anti-*S. faecalis* aldolase serum. Comparisons of aldolases from organisms *within* a given antigenic group invariably produced a pattern of identity or apparent identity (confluent lines of precipitation), whereas comparisons *between* groups produced patterns of partial identity (single spurred precipitates). Because the protein marker was purified from a Lancefield group-D *Streptococcus,* the enterococci automatically became the reference point for all comparisons. The seven antigenic groups shown in Fig. 8 are arranged in descending order of their relatedness to the reference enzyme. The absence of crossed or double-spurred precipitates signifies that the streptococcal aldolases have not undergone detectable antigenic divergence. Such findings can be interpreted to mean that, *with respect to the aldolase marker,* the streptococci have evolved in a relatively linear or unidirectional fashion. However, when other proteins from organisms within each of the groups become available, the conclusions drawn from the aldolase study may require some modification. The summary presented in Fig. 8 clearly indicates that the species of streptococci surveyed are related through some common ancestor. However, the appearance of representatives of two species, *Streptococcus mutans* and *Streptococcus mitis,* in more than one antigenic group raises serious questions about the homogeneity of either taxon and their recent ancestry.

It is somewhat more difficult to arrange the lactobacilli in the same fashion as the streptococci, because several aldolases produce double-spurred precipitates on cross-matching; however, the descending order of relatedness and the divergent antigenic outcroppings are shown in Fig. 9. The genus *Lactobacillus* contains at least 10 distinct aldolase antigenic groups. It is of interest to note that the aldolases that produce double spurs belong to the distinctive 76,000 molecular-weight class represented by *L. xylosus,* *L. salivarius,* and *L. plantarum.* Again, strains assigned to a single species, in this instance *Lactobacillus acidophilus,* appear in two different antigenic groups. Gasser and Gasser (17) have suggested that the species *L. acidophilus* contains as many as four distinct biotypes and is, therefore, a heterogeneous collection of lactobacilli. The pediococcal aldolases exhibit the least antigenic diversity of the three genera examined and can be assigned to one of three antigen groups (Fig. 9). Like the streptococci, only reactions of apparent identity and partial identity are observed in cross-matches, which suggests that the evolution of the genus has been unidirectional. Ironically, the *Pediococcus parvulus* aldolase, whose physical properties most resemble those of the reference enzyme, is the least immunologically related to it.

The intergeneric cross-matches were perhaps the most illuminating immunodiffusion

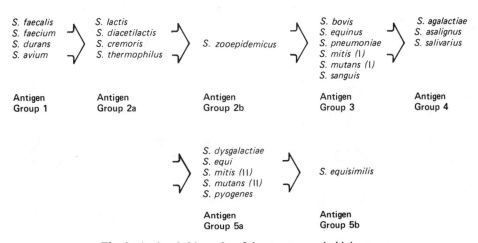

Fig. 8. Antigenic hierarchy of the streptococcal aldolases.

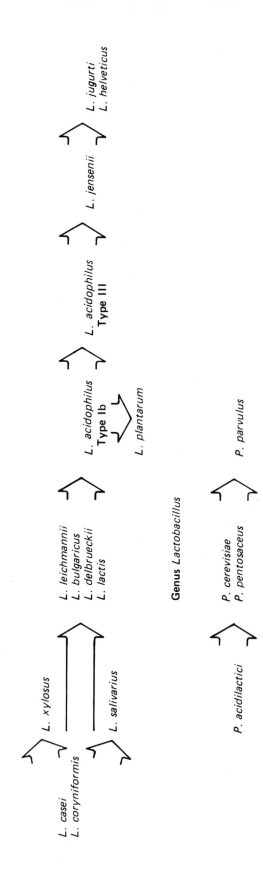

Fig. 9. Order of immunologic relatedness of lactobacilli and pediococcal aldolases.

L. jugurti
L. helveticus

L. jensenii

L. acidophilus
Type III

L. acidophilus
Type Ib

L. plantarum

P. parvulus

L. leichmannii
L. bulgaricus
L. delbrueckii
L. lactis

Genus Lactobacillus

P. cerevisiae
P. pentosaceus

Genus Pediococcus

L. xylosus

L. salivarius

L. casei
L. coryniformis

P. acidilactici

72

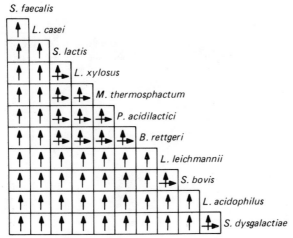

Fig. 10. Inter- and intrageneric cross-matches of lactic acid bacteria aldolases. Arrows indicate dominant antigen in pairwise cross-match. Crossed arrows indicate reactions of nonidentity.

experiments of this series. These experiments are summarized in Fig. 10 according to the convention of Gasser and Gasser (17), in which the arrow indicates the dominant antigen in the cross-match. Although the malic enzyme experiments established a "natural" bridge between *S. faecalis* and *L. casei*, the observation that the aldolase of the latter is the most closely related isolog to *S. faecalis* aldolase was unexpected. The double-spurred precipitates or reactions of nonidentity produced by extensive pairwise cross-matching of the aldolases from *Streptococcus lactis*, *Pediococcus acidilactici*, *Microbacterium thermosphactum*, *L. xylosus*, and *Butyribacterium rettgeri* indicate that each of the species in this diverse group of microorganisms is developing along distinct and separate lines from one another. The divergence between the streptococci and lactobacilli is confirmed by cross-matches between the more immunologically remote members of either genus. For example, *Streptococcus bovis* and *Lactobacillus leichmannii* produce cross-spurs against each other, as do *Streptococcus dysgalactiae* and *L. acidophilus* (Fig. 10). The same Figure also shows the antigenic hierarchies that exist within and between groups.

During the immunologic survey of the streptococcal, pediococcal, and *Lactobacillus* aldolases, the aldolases from several bacteria outside of the tribe Lactobacilleae were included in the testing to provide a boundary that would serve to delineate the related forms (15). The inclusion of the *B. rettgeri* aldolase in the survey produced an unexpected cross reaction with anti-*S. faecalis* aldolase serum that eventually led to a survey of other anaerobic bacteria. Species of gram-positive nonspore-forming anaerobic bacteria from the genera *Eubacterium*, *Propionibacterium*, and *Arachnia* were also tested for aldolase cross reactivity. As before, immunodiffusion experiments were used to establish the antigenic hierarchy of this group (Fig. 11). The *B. rettgeri* aldolase bore the greatest similarity to the *S. faecalis* enzyme, while the propionibacteria were the least related. *Arachnia propionica* extracts only produced a weak, hazy precipitate upon interacting with the antiserum.

Table 5. Immunologic Distances of Streptococcal Aldolase Using *S. faecalis* Aldolase as the Reference Protein

Organism	Strain	Cell wall antigen	Index of dissimilarity	Immunologic distance
Antigen group 1 [a]				
S. faecalis	MR (homologous)	D[c]	1.0	0
S. faecalis	N83	D	1.1	4
S. faecium	K6A	D	1.0	0
S. faecium	N55	D	2.1	32
S. faecalis var. liquefaciens	835-69	D	1.06	2.5
S. faecalis var. liquefaciens	1773-71	D	1.51	8
S. avium	SS-559	Q	2.6	41
S. durans	ATCC 19432	D	2.7	43
Antigen group 2				
S. lactis	ATCC 19435	N	8.3	92
S. cremoris	ATCC 19257	N	9.0	95
S. thermophilus	ATCC 19258	—	10.4	102
S. diacetilactis	ATCC 11007	N	13.5	113
Antigen group 2b				
S. mitis	ATCC 15909	I[d]	18	125
Streptococcus sp.	SS-665	K	22	134
S. mitis	ATCC 15912	IV	25	139
S. zooepidemicus	SS-259	C	27	143
Antigen group 3				
S. bovis (variant)	27-72	—	30	147
S. bovis	ATCC 9809	D	34	153
S. bovis	ATCC 15351	D	39	159
S. mitis	ATCC 15911	III	33	152
S. mutans				
	SL-1	d[e]	35	154
	Bergergren	d	41	161
	6715	d	42	162
	01H1	d	45	165
S. sanguis				
	Wicky SS-975	H	30	147
	F90A SS-546	H	32	150

Neimark (48,49) has recently demonstrated that the anti-*S. faecalis* aldolase serum also reacts with extracts of the mycoplasmas that do not require sterols. They were shown to be related in the descending order of *Acholeplasma laidlawii* > *Acholeplasma xanthum* > *Acholeplasma granularum*. Double-spurred reactions occurred between *A. laidlawii* and *S. lactis*.

Fig. 11. Antigenic hierarchy of the aldolases of some gram-positive nonspore-forming anaerobic bacteria.

Table 5. *Continued*

Organism	Strain	Cell wall antigen	Index of dissimilarity	Immunologic distance
	Channon SS-982	H	35	154
	Blackburn SS-983	H	35	154
	ATCC-10556	H	39	159
	K208 SS-547	H	40	160
S. equinus	ATCC 9812	D	39	159
D. pneumoniae	ATCC 6308	—	39	159
Antigen group 4				
Streptococcus sp.	SS-740	U	49	169
S. asalignus	ATCC 8059	B	56	175
Streptococcus sp.	ATCC 8144	H	61	178
S. salivarius	ATCC 13813	B	66	182
S. salivarius	112	K	66	182
S. salivarius	9222	K	66	182
	ATCC 13419	K	70	184
Antigen group 5				
S. mitis				
	ATCC 15914	VII	70	184
	ATCC 9895	V	80	190
	ATCC 15910	II	88	194
S. pyogenes	ATCC 14298	A	80	190
S. equi	ATCC 6580	C	81	190
Streptococcus sp.	ATCC 9932	L	82	191
S. dysgalactiae	ATCC 9926	C	88	194
S. mutans	ATCC 10449	c	85	193
	JC-2	c	85	192
	BHT	b	89	194
	FA-1	b	109	203
	E-49	a	131	211
Antigen group 5b				
S. equisimilis	ATCC 9542	C	105	202

[a] Antigen groupings from immunodiffusion cross-matches.

[b] For source of organisms, see references.

[c] Capital letters designate Lancefield-group antigens.

[d] Roman numerals designate Williamson serotypes.

[e] Lower case letters designate Bratthall serotypes.

The immunodiffusion experiments established a rough ordering of the aldolases found in representatives of the following genera: *Streptococcus, Lactobacillus, Pediococcus, Microbacterium, Butyribacterium, Eubacterium, Propionibacterium, Arachnia,* and *Acholeplasma.* The findings also indicate that there exist at least six diverging lines of aldolase evolution. However, microcomplement fixation experiments were required for the accurate placement of the respective species within each of the evolutionary lines. The differences in the kind of information each of these procedures yield are discussed elsewhere (15). Tables 5 and 6 summarize the results obtained from a comprehensive comparison of the various aldolases with the use of anti-*S. faecalis* MR aldolase serum. The marked increase in sensitivity of the microcomplement fixation technique over immunodiffusion is immediately apparent from the streptococcal aldolase data. A significant degree of antigenic divergence was observed within each of the antigenic groups by

Table 6. Immunologic Distances of Lactobacilli, Pediococci, and Other Related Organisms Using the *S. faecalis* Aldolase as the Reference Protein

Genus *Lactobacillus* species and strain		ID[a]	Immunologic distance	Gram-positive nonspore-forming anaerobic bacteria		
				Species	ID	Immunologic distance
L. coryniformis sp. *torquens*	M30	3.5	54	*B. rettgeri* 10852	7.75	89
L. coryniformis sp. *coryniformis*	M34	3.6	56	*E. limosum* 8486	8.2	91
L. casei sp. *pseudoplantarum*	M40	4.7	67	*E. cylindroides* (3594)	33	151
L. zeae	15820	4.7	67	*E. cylindroides* (3696)	34	153
L. casei sp. *rhamnosus*	F.3.4	4.7	67	*E. aerofaciens*-25986	108	203
L. casei sp. *casei*	64H	4.8	68	*P. jensenii* (4868)	148	217
L. casei sp. *rhamnosus*	F.3.3	5.0	70	*P. zeae* 4964	185	226
L. casei sp. *rhamnosus*	OC91	5.4	73	*P. intermedium* 14072	190	227
L. casei sp. *alactosus*	C45	5.8	76	*P. acnes* 6919	191	228
L. casei sp. *rhamnosus*	CL11	6.3	80	*P. peterssonii* 4870	212	232
L. xylosus	15577	20	130	*P. pentosaceum* 4875	215	233
L. salivarius	11741	30	148	*A. propionica* 14157	>220	
L. bulgaricus	11842	36	155			
L. lactis	12315	39	159			
L. delbrueckii	9649	41	161			

L. leichmannii	4797	43	163
L. curvatus	M1	116	206
L. mali	27053	133	212
L. plantarum	14917	138	214
L. acidophilus(Ib)	19992	176	224
L. acidophilus(III)	4356	203	230
L. jensenii	25258	225	235
L. jugurti	521	268	243
L. helveticus	15009	290	246

Genus *Pediococcus*
Species

P. acidilactici	25740	9.2	96
P. acidilactici	25742	11.2	104
P. cerevisiae	559	19.5	129
P. cerevisiae	8042	19.8	129
P. pentosaceus	25744	23	136
P. cerevisiae	990	20	130
P. parvulus	13371	39	159

Others

Microbacterium thermosphactum	11509	11.7	106

[a] Index of dissimilarity.

microcomplement fixation. The diversification that occurred among the enterococci is most obvious, because organisms that belong to this group share a greater number of antigenic determinants with the reference *S. faecalis* MR aldolase than do the other streptococci. It is also clear that as the immunologic distances between the homologous and heterologous antigens increase, only those determinants located in conserved regions of the polypeptide chain will remain relatively unaltered. For this reason, the actual extent of divergence within the more remote antigen groups cannot be fully appreciated. Without internal markers for each of the aldolase antigen groupings, a complete understanding of the various interspecific relationships among the immunologically remote species is not possible. The quantitative serologic studies are concordant with the immunodiffusion experiments and support the notion that the species *S. mutans* and *S. mitis* are a heterogeneous collection of streptococci. Bratthall (50) and Williamson (51) have previously demonstrated the presence of distinctive cell wall serotypes among strains of *S. mutans* and *S. mitis,* respectively, and Coykendall (52), with DNA hybridization studies, has demonstrated the existence of four distinct genotypes within the species *S. mutans.*

The species of *Lactobacillus* appear in an order of descending similarity, which corresponds to that obtained with the immunodiffusion experiments (compare Fig. 9 with Table 6). In this genus, the results of the aldolase study can be compared with data reported for a lactate dehydrogenase (LDH) study by Gasser and Gasser (17). Although antiserum prepared against the L-LDH of *L. acidophilus* reacted only weakly with the LDHs of *L. casei* and *L. salivarius,* good reactivity was observed with the other homofermentative lactobacilli tested. Both enzyme-antibody systems recognized the same groups of apparent identity, namely, the *L. leichmannii* group (see Fig. 9) and the *Lactobacillus jugurti* group. The ordering of the species by the two protein markers could not be directly compared, because the anti-LDH serum detected a bifurcation between strains of *L. acidophilus* that was not discerned by the antialdolase serum. The discrepancy in the two sets of results can be attributed to the relatively great immunologic distance between the aldolase reference organism and the *L. acidophilus* aldolases and points out the importance of selecting other well-placed group markers. However, if the two diverging lines are arranged along a single axis, the ordering produced by the aldolase system very closely resembles that of the LDH system.

5.3   Natural Relations of the Lactic Acid Bacteria Based on a Second Reference Aldolase Marker

In any immunologically based phylogenetic study, it is always advisable to confirm the conclusions derived from the study of one protein marker with a study of a second related protein. It was particularly essential in the case of FDP aldolase because the variation in the quaternary structure of the enzymes from lactic acid bacteria raised the possibility that exposed antigenic determinants on the surface of the smaller forms of the enzyme might be masked within the larger enzyme aggregates. Such a masking could certainly alter the immunologic relationships between the various size classes of aldolase, with the larger forms of the enzyme appearing less related to the smaller ones. Thus, the second FDP aldolase selected for this study came from *P; cerevisiae*; the 176,000 molecular-weight enzyme appeared to be a typical representative of the largest size class of aldolase present among these bacteria.

Reciprocal experiments similar to those employed in the malic enzyme study were used to confirm and extend the earlier results. While the titers of the antipediococcal aldolase sera were significantly lower than those of the antistreptococcal aldolase sera and limited the scope of the survey, the reactivity of the former was sufficiently great to confirm a large portion of the original experiments. In the original study, the four species within the genus *Pediococcus* were segregated into three distinct antigenic groups (see Fig. 9). The antipediococcal aldolase serum immunodiffusion experiments produced the same pattern of clustering in the order *P. cerevisiae* = *Pediococcus pentosaceus* > *P. acidilactici* > *P. parvulus*. Aldolases of the Lancefield group-D strains of *Streptococcus* exhibited a greater degree of antigenic similarity to pediococcal enzyme than did the Lancefield group-N (lactic) streptococcal aldolases. Aldolases of the remaining three major antigenic groupings reacted too weakly with antipediococcal aldolase serum to be of use in comparative cross-matches. The majority of the lactobacilli aldolases reacted quite strongly with the antipediococcal aldolase serum, and pairwise cross-matches produced the following decreasing order of antigenic similarity: *Lactobacillus coryniformis* ≧ *L. casei* > *L. salivarius* > *L. leichmannii* = *L. delbrueckii* = *L. lactis* = *L. bulgaricus* > *L. acidophilus*. Cross reactions with the more antigenically remote forms of *Lactobacillus* aldolase were too weak to be of use in immunodiffusion experiments. However, the ordering of the above species corresponds precisely to that described in the previous work (15). The previously untested aldolase of *Aerococcus viridans* reacted with antipediococcal aldolase serum but failed to react with the antistreptococcal aldolase serum, indicating that the alterations in the primary structure of the *Aerococcus* enzyme had exceeded the limits of detection of the latter antiserum.

Interspecific cross-matches between the lactobacilli and streptococci revealed that, with one possible exception, the pediococcal aldolases are more closely related to the

Table 7. Comparison of Immunologic Distances Obtained with Antipediococcal and Antistreptococcal aldolase

Organism	Immunologic distance	
	Antipediococcal aldolase	Antistreptococcal aldolase
P. cerevisiae 559	0	129
P. cerevisiae 990	5	130
Pediococcus pentosaceus ATCC 25744	8	126
P. acidilactici ATCC 25740	21	96
P. parvulus ATCC 19371	121	159
Aerococcus viridans ATCC 11563	152	No reaction
S. faecalis MR	171	0
Streptococcus lactis ATCC 19435	175	91
L. casei 64H	73	68
L. bulgaricus ATCC 11842	145	158
L. salivarius ATCC 11741	160	147
L. acidophilus ATCC 19992	176	224
M. thermosphactum ATCC 11509	170	106
B. rettgeri ATCC 10825	162	89
Propionibacterium pentosaceum ATCC 4875	139	232
Propionibacterium arabinosum ATCC 4965	144	230

aldolases of certain lactobacilli than to the streptococcal aldolases; *P. parvulus* does not appear to fall into the main cluster of pediococci and thus may be antigenically distant from the lactobacilli. The close relationship between the pediococci and lactobacilli was predictable from the similarities in the physical properties of their aldolases. Both the *L. casei* and *P. cerevisiae* enzymes are aggregates of a 29,500 molecular-weight subunit, in contrast to the dimeric *S. faecalis* aldolase, which is composed of two 28,000 molecular-weight monomers.

Microcomplement fixation assays performed with antipediococcal aldolase serum now served to "fix" the positions of several lactic acid bacteria into a two-dimensional plane by providing certain bacterial species with two reference points (indices of dissimilarity). The immunologic distances for aldolases of key lactic acid bacteria are presented in Table 7. These results confirm the immunodiffusion data and provide a precise placement for the respective bacteria; they are also presented with the immunologic distances from the antistreptococcal aldolase study for comparative purposes. The extent of relatedness between the *L. casei* group and the pediococci is immediately apparent. Also, there exists a greater antigenic similarity between the aldolases of the propionibacteria and pediococci than between the propionibacteria and streptococci. With the composite data, a tentative phylogenetic map of the lactic acid bacteria and related organisms was prepared.

6. A PHYLOGENETIC REPRESENTATION OF THE HOMOFERMENTATIVE LACTIC ACID BACTERIA BASED ON IMMUNOLOGIC SIMILARITIES OF THEIR ALDOLASES

A phenogram that depicts the antigenic relationships found among lactic acid bacteria aldolases was far more difficult to prepare than was the malic enzyme representation, for several reasons. First, the latter study dealt only with two species of bacteria, while the aldolase work attempted to draw relationships among some 50 species. Second, the malic enzymes exhibited no major antigenic divergences, whereas the FDP aldolase appears to have undergone extensive evolutionary diversification. Finally, the malic enzyme exists in a single quaternary form, in contrast to the five distinct quaternary forms of aldolase. It is this last fact that is particularly troublesome when attempting to develop a natural system of classification for this large group of organisms based on FDP aldolase, because, at the present, there are no means by which the functional and evolutionary impacts of the structural changes can be evaluated. Nor is there any way to estimate the number of amino acid substitutions that would be required to produce the variations in quaternary structure. Therefore, in the initial report of this work (15), an arbitrary decision was made to distinguish between the various molecular weight classes of aldolase when they occurred within a given branch of the map and were not the major form found therein. The continuation of this convention here serves no other purpose than to point out the position of the less common forms of aldolase and does not imply that the data can be used to distinguish between structural changes that occur as a result of Darwinian selection as opposed to alterations produced by neutral or silent mutations (53).

The phenogram shown in Fig. 12 was first prepared in a rough skeletal form drawing solely from the anti-*S. faecalis* aldolase immunodiffusion results. The six distinct evolu-

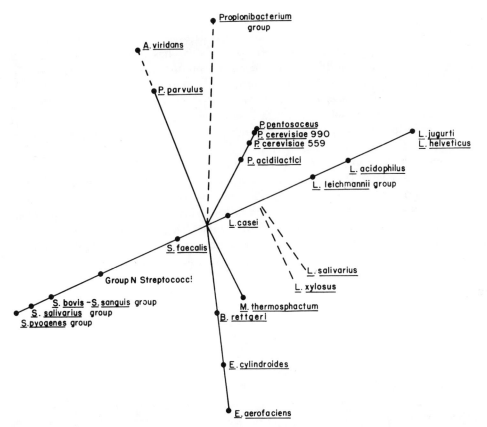

Fig. 12. A phylogenetic map of the lactic acid bacteria and related organisms based on the immunologic similarities that exist among their aldolases.

tionary lines were mandated by the double-spurred cross-matches (reactions of nonidentity) that occurred in this set of data, while the order within each line was established from the single-spurred cross-matches (reactions of partial identity). A precise placement of the individual species in the various lines was achieved by defining an immunologic distance unit as an arbitrary unit of measurement (1 mm in the original drawing). Because the immunodiffusion results obtained with the second marker, the *P. cerevisiae* aldolase, neither altered nor conflicted with the placement of the various species, as determined from the original data, their positions were adjusted until they fit both sets of immunologic distances. The excellent agreement in the two sets suggest that there is little or no masking of antigenic determinants in the hexameric form of the aldolase.

In the original map (15), all of the evolutionary lines radiated outward from the *S. faecalis* locus, because it automatically became the sole point of reference. The inclusion of the second marker rearranged several species and resulted in a shift of the pediococcal branch and the anaerobe branch; these branches now intersect the streptococci and lactobacilli at a point between the two genera. The other groups, which are represented by a single locus on the map, were arbitrarily joined at the same point. It is very tempting to speculate that the point of intersection may represent the common ancestor (which may or may not be extant); however, considerably more information must be gathered before such a hypothesis can be seriously discussed. The second map differs from the original version in several other respects; *P. parvulus* has been completely separated from the major cluster of pediococci, which suggests that its genesis may represent an independent evolutionary event that is unrelated to the development of the rest of the genus. Similarly, *L. salivarius* and *L. xylosus* map well away from the main *Lactobacillus* branch; however, sufficient data are not yet available to decide whether these two species arose from a contemporary form within the branch or from some more ancient progenitor not shown on the map.

The map depicted in Fig. 12 would be significantly enlarged if the results of the LDH study by Gasser and Gasser (17) were incorporated into it. Their mapping shows that the heterofermentative lactobacilli and *Leuconostoc* species extend outward and away from the homofermenters to form a second large constellation of the lactic acid bacteria. Although a precise integration of the two studies cannot be attempted until it is established that both sets of immunologic distances are equivalent, the agreement in the overlapping areas does indicate that the two maps are contiguous with regard to the ordering of species in this large related group of microorganisms.

7. MOLECULAR EVOLUTION OF THE LACTIC ACID BACTERIA ALDOLASE

From the immunologic cross reactivity observed among the different molecular weight classes of aldolase (15), and from the similarities in the size of their subunits (29), it appears that a single, relatively conserved subunit has undergone subtle alterations in its primary structure from which a variety of size aggregates have arisen. It was apparent from our earlier work (15) that the vast majority of the aldolases exist as aggregations of even numbered subunits, namely, 2, 4, and 6. In fact, only three species of lactic acid bacteria appear to possess the trimeric form of the enzyme, and only a single species contains the pentameric form of the enzyme.

If the various molecular weight classes of aldolase do, indeed, represent different quaternary arrangements of a structurally conserved monomer, it is not difficult to envisage how relatively minor amino acid substitutions altered the aggregation patterns of these enzymes during the course of their evolution. The discrepancy in size between the 28,000 molecular-weight streptococcal aldolase subunit and the 29,500 molecular-weight aldolase subunit of the pediococci and lactobacilli is roughly equivalent to a difference of 10 amino acid residues, assuming a molecular weight of 150 per amino acid. Because at least two binding sites are required to form an aggregate of three or more protomers, the size difference may explain why the streptococcal enzyme exists, almost exclusively, as a dimeric protein. If, during the early stages of streptococcal evolution, a deletion mutation occurred in the portion of the aldolase gene that coded for one of the two subunit interaction sites required for oligomer formation, a dimer would be the largest possible aggregate. It follows, therefore, that the tetrameric form of the enzyme probably differs from the hexameric form in the location of the two binding sites.

Without paleontologic records or amino acid sequence analyses, there is no way to accurately determine which of the five aldolase configurations is most similar to the parent, or ancestral, enzyme. However, the immunologic and structural studies do provide some clues as to the ancestry of the enzyme. For example, the hexameric forms of aldolase found among the pediococci and propionibacteria are probably of relatively recent origin, because they exhibit the least antigenic divergence. This notion is supported, in part, by the work of Johnson (54), who established that the propionibacteria are still a relatively homogeneous genetic group; should DNA hybridization studies among the pediococci produce a similar result, the accuracy of these protein homology studies will be further borne out. While the streptococcal aldolases show extensive immunologic diversification, the diversity among the aldolases of the lactobacilli is even greater, with a range of indices of dissimilarity that exceeds that of the streptococcal aldolases.

The modification in the size of the streptococcal aldolase subunit also appears to support the argument for "a more recent origin" of this group of organisms, because the loss of a polypeptide segment in a protein molecule is a more likely event than the acquisition of such a segment. Thus, the limited amount of indirect evidence presently at hand indicates that the parental form of the FDP aldolase found among lactic acid bacteria was probably a tetramer.

This tentative conclusion raises several intriguing questions about the genesis of the other forms of aldolase. For example, if the size of the *S. salivarius* aldolase subunits are similar to those of the *S. faecalis* enzyme, how was the primary sequence of the subunit modified to produce a new second binding site and the resultant pentameric aggregate? Do the trimeric forms of aldolase found in three *Lactobacillus* species represent a recent modification of the tetrameric form of the enzyme? And what, if any, selective advantages are derived from the modifications that produced the trimeric and pentameric forms of aldolase? Such questions must await a more detailed study of the structure of these enzymes.

8. REEVALUATING SEVERAL SHIBBOLETHS OF CLASSIC MICROBIAL TAXONOMY

8.1 Morphologic Diversity and Protein Homology

Cohn (55,56), one of the pioneers of bacterial taxonomy, recognized both the advantages and dangers in using morphologic characteristics to classify prokaryotes. Although contemporary taxonomists make liberal use of the large body of nutritional and physiologic information that has been amassed since the work of the early systematists, morphologic considerations still maintain a prime position in the most elaborate and sophisticated classification schemes. This practice continues despite explicit warnings regarding its pitfalls (57,58). Obviously, some degree of importance can be attributed to the morphologic differences that exist among the three major groups of homofermentative lactic acid bacteria, because a rough correlation is observed between the various biologic forms and the size of their aldolases. The data suggest that the mutational events that produced a specific shape, i.e., streptococcus, streptobacillus, and tetrad (coccus), occurred concomitantly with a modification in the size of the aldolase subunit aggregate. However, in many instances, the similarity in size of the aldolases bears no relationship to the degree of structural homology between the respective enzymes. The evidence for this is as follows: (i) immunologically, the 56,000 molecular-weight *S. faecalis* aldolase is more closely related to the 121,000 molecular-weight *L. casei* aldolase than to its nearest Lancefield group-N streptococcal neighbor; (ii) the degree of antigenic homology between the group-N streptococcal aldolases and the *S. faecalis* aldolase is no greater than that observed between the latter and *M. thermosphactum*, *B. rettgeri*, and *P. acidilactici*; and (iii) the pediococcal aldolase is more closely related to the *L. casei* enzyme than to the aldolase in its spherical counterpart *S. faecalis*. If the immunologic relationships of the FDP aldolase accurately reflect natural affinities among the lactic acid bacteria, it is apparent that morphologic traits cannot always be relied on to point up phylogenetic relationships among bacteria. Moreover, in addition to the comparatively close immunologic relationships that exist between the pediococcal and *Lactobacillus* aldolases, the aldolase subunits of these two forms also share a greater degree of physical similarity than that found between the monomers of pediococcus and streptococcus. The combined findings suggest that the two coccal forms, streptococcus and pediococcus, may have arisen from a rod-shaped progenitor as independent evolutionary events and that one coccal form did not give rise to the other.

The problems posed for classic systematics by protein relatedness studies are probably best exemplified by examining the manner in which the lactic acid bacteria have been classified in the recently published eighth edition of *Bergey's Manual of Determinative Bacteriology* (59). The lactic acid bacteria are segregated according to shape, the rod-shaped forms were assigned to the family Lactobacilleaceae, while the spherical organisms were placed in the family Streptococcaceae. The latter family contains the genera *Pediococcus* and *Leuconostoc*, both of which have been shown to be more closely related to the lactobacilli (15,17) than the streptococci. Examination of the LDH map (17) and the aldolase map, on one hand, and Bergey's manual, on the other, suggests that the primary division of the certain microbial groups based on present morphologic differences may not be consistent with their evolutionary development.

8.2   Phenotypic Convergence amid Phylogenetic Divergence

Biochemical and nutritional properties are usually reliable phenotypic characters that provide a sound basis for relating or differentiating bacterial taxa, especially among those groups of microorganisms that possess numerous and diverse biochemical activities. However, there are several examples among the lactic acid bacteria that provide exceptions to this rule; in these instances, genotypic differences give false impressions of phenotypic similarities. During their LDH study, Gasser and coworkers (60,61) isolated and characterized several *Lactobacillus* strains that were phenotypically identical to *L. leichmannii*. Because the two groups of organisms could readily be distinguished from one another by simply comparing certain physical properties of their LDHs or DNA base ratios, a new species, *Lactobacillus jensenii*, was created to accommodate the new isolates. Gasser and Gasser (17) later referred to such phenotypic look-alikes as "sosie" species.

At least two similar situations exist among the streptococci, and both of them become apparent when the immunologic data are presented as a phenogram (Fig. 13), prepared according to the convention of Cocks and Wilson (11). *S. mutans* is a collection of phenotypically similar organisms that comprise at least five distinct serotypes (50) and four discernible genotypes (52). The comparative immunologic studies with aldolase indicate that one of the genotypes *may not even share recent ancestry* with the three remaining groups (Fig. 13). With six serotypic subdivisions (51), the species *S. mitis* also appears to be a heterogeneous group of streptococci, and the appearance of these serotypes at multiple loci in the phenogram (Fig. 13) supports this contention. While it

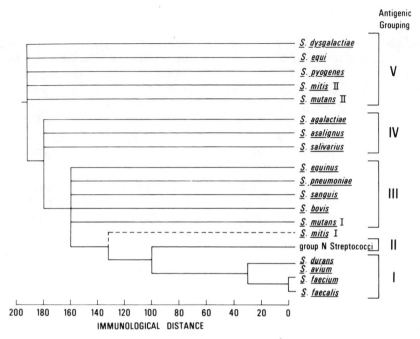

Fig. 13. A phenogram of the genus *Streptococcus* that represents the antigenic relationships found among the streptococcal aldolases.

is conceivable that a single phenotype might be imposed on several genotypically distinct groups (50) of bacteria by chance, it is far more likely that such an occurrence is the result of selective pressures. Because the four genotypic groups of *S. mutans* must compete with one another and with other streptococci in a highly specialized environment, the oral cavity, their specific phenotype may confer upon them the ability to survive in this particular econiche. Alternatively, a highly successful phenotype might be conserved by conserving the enzymatic protein components that define it and thereby evolve as slowly as the aldolase marker; however, there is no evidence that this is the case here.

It should be pointed out that there exist phenotypically differentiable lactobacilli that are genotypically similar or identical. The four species, *L. leichmannii, L. delbrueckii, L. lactis,* and *L. bulgaricus,* were originally defined and segregated by certain distinctive phenotypic characteristics. Gasser and Gasser (17) were the first to cluster these four species under the heading of the *L. leichmannii* group based on reactions of immunologic identity in the LDH : anti-LDH system. The same group of identical specificity was observed with the aldolase : antialdolase system, which suggested that the four species were, indeed, closely related (15). Subsequently, DNA hybridization studies by Sriranganathan et al. (62) established that the four species share so high a degree of homology that they probably represent a single genetic group. From these examples, it is apparent that biochemical characteristics rarely provide an absolute degree of reliability in microbial classifications. The substrates and products of enzymatic reactions may indirectly reflect the evolutionary history of an organism, but only the catalytic agents, the enzymes, bear the imprint of the bacterium's genetic history.

9. CONCLUDING REMARKS

The studies with cytochromes (63), globins (64), and lysozymes (65,66) leave little doubt that protein homology studies can be used to develop natural systems of classification for higher organisms. The very large extant literature on paleontologic and comparative anatomic research has benefited these studies by providing geologic time frames for the more biologically complex forms of life.

These reference points in time permit biochemical taxonomists to create phylogenetic maps with their primary amino acid sequence analysis data that reflect cladistic relationships of eukaryotic organisms. With only a few exceptions, identifiable fossil records of prokaryotes do not exist; fossil imprints thought to be blue-green bacteria have been found in sandstone and iron-bearing deposits (67,68). The absence of both paleontonlogic records and complex developmental histories has been the major obstruction to the development of natural classifications for bacteria. Indeed, even if extensive sequence analysis data were available for sets of prokaryotic proteins, detailed phylogenetic maps would still lack time scales and, as such, only depict phenetic rather than cladistic relationships. Additional constraints are placed on the development of natural classifications when protein homologies are established by serologic techniques instead of amino acid sequencing. While Prager and Wilson (35,69) demonstrated that it is possible to correlate the *number* of amino acid substitutions that occur in a set of structurally related proteins with the quantitative immunologic differences observed between them, the procedures cannot identify *specific* amino acid substitutions. Furthermore, serologic procedures appear to be restricted to detecting primary sequence substitutions that range from only 28 to 40% of the total amino acid complement (69); alterations that exceed this limit result in the loss of immunologic cross reactivity. Fortunately, the formerly held notion that immunologic methods only detect substitutions on the external portions of protein molecules appears to be erroneous, because Reichlin (70) and Prager et al. (71) demonstrated that amino acid substitutions within the conserved, buried regions of a protein can be expressed as conformational alterations on the surface of the molecule that produce different antigenic determinants. However, even within these somewhat narrower confines, sensitive serologic techniques can be used to measure the effects of evolutionary pressures on the structure of protein molecules.

ACKNOWLEDGMENTS

The excellent technical assistance of Ms. E. Y. Meyer, K. Kline, S. Kulczyk, and N. M. Chace is greatly appreciated.

REFERENCES

1. M. R. Bethell, R. von Fellenberg, M. E. Jones, and L. Levine, *Biochemistry,* **7,** 4315 (1968).
2. R. von Fellenberg, M. R. Levine, M. E. Jones, and L. Levine, *Biochemistry,* **7,** 4322 (1968).
3. L. Levine and H. Van Vunakis, "Micro-Complement Fixation," in C. W. H. Hirs, Ed., *Methods in Enzymology,* Vol. XI, Academic Press, New York and London, 1967, p. 928.

4. M. Boll, F. Falkenberg, D. Jeckel, and G. Pfleiderer, *Z. Physiol. Chem.*, **351**, 1268 (1970).
5. R. Pietruszko and H. J. Ringold, *Biochem. Biophys. Res. Commun.*, **33**, 497 (1968).
6. D. Perrin, *Ann. N.Y. Acad. Sci.*, **103**, 1058 (1963).
7. A. V. Fowler and I. Zabin, *J. Mol. Biol.*, **33**, 35 (1968).
8. E. M. Prager and A. C. Wilson, *J. Biol. Chem.*, **240**, 7010 (1971).
9. R. N. Patel, W. J. Mandy, and D. S. Hoare, *J. Bacteriol.*, **113**, 937 (1973).
10. A. C. Wilson and N. O. Kaplan, "Enzyme Structure and Its Relation to Taxonomy," in C. A. Leone, Ed., *Taxonomic Biochemistry and Serology*, Ronald Press, New York, 1964, p. 321.
11. G. T. Cocks and A. C. Wilson, *J. Bacteriol.* **110**, 793 (1972).
12. D. L. Steffen, G. T. Cocks, and A. C. Wilson, *J. Bacteriol.*, **110**, 803 (1972).
13. T. L. Whiteside, A. J. DeSiervo, and M. R. J. Salton, *J. Bacteriol.*, **105**, 957 (1971).
14. S. W. Tafler, P. Setlow, and L. Levine, *J. Bacteriol.*, **113**, 18 (1973).
15. J. London and K. Kline, *Bacteriol. Rev.*, **37**, 453 (1973).
16. S. R. Tronick, J. E. Ciardi, and E. R. Stadtman, *J. Bacteriol.*, **115**, 858 (1973).
17. F. Gasser and C. Gasser, *J. Bacteriol.*, **106**, 113 (1971).
18. J. London, E. Y. Meyer, and S. Kulczyk, *J. Bacteriol.*, **106**, 126 (1971).
19. J. London, E. Y. Meyer, and S. R. Kulczyk, *J. Bacteriol.*, **108**, 196 (1971).
20. R. Y. Stanier, D. Wachter, C. Gasser, and A. C. Wilson, *J. Bacteriol.*, **102**, 351 (1970).
21. L. Bogorad, *Rec. Chem. Progr.*, **26**, 1 (1965).
22. D. S. Berns, *Plant Physiol.*, **42**, 1569 (1967).
23. A. N. Glazer, G. Cohen-Bazire, and R. Y. Stanier, *Proc. Nat. Acad. Sci. USA*, **68**, 3005 (1971).
24. B. A. McFadden and A. R. Denend, *J. Bacteriol.*, **110**, 633 (1972).
25. T. E. Creighton, D. R. Helinski, R. L. Somerville, and C. Yanofsky, *J. Bacteriol.*, **91**, 1819 (1966).
26. T. M. Murphy and S. E. Mills, *J. Bacteriol.*, **97**, 1310 (1969).
27. V. Rocha, I. P. Crawford, and S. E. Mills, *J. Bacteriol.*, **111**, 163 (1972).
28. J. London and E. Y. Meyer, *J. Bacteriol.*, **98**, 705 (1969).
29. J. London, *J. Biol. Chem.*, **249**, 7977 (1974).
30. O. Ouchterlony, *Handbook of Immunodiffusion and Immunoelectrophoresis*, Ann Arbor-Humphrey Science Publ., Ann Arbor-London, 1968.
31. D. Stollar and L. Levine, in S. P. Colowick and N. O. Kaplan, Eds., *Methods in Enzymology*, Vol. VI, Academic Press, New York and London, 1963, p. 848.
32. E. Wasserman and L. Levine, *J. Immunol.*, **87**, 290 (1961).
33. A. B. Champion, E. M. Prager, D. Wachter, and A. C. Wilson, "Microcomplement Fixation," in C. A. Wright, Ed., *Biochemical and Immunological Taxonomy of Animals*, Academic Press, New York and London, 1974, p. 397.
34. V. M. Sarich and A. C. Wilson, *Science*, **154**, 1563 (1966).
35. E. M. Prager and A. C. Wilson, *J. Biol. Chem.*, **246**, 5978 (1971).
36. W. Rutter, *Fed. Proc.*, **23**, 1248 (1964).
37. J. London and E. Y. Meyer, *Biochim. Biophys. Acta*, **178**, 205 (1969).
38. S. Ochoa, A. H. Mehler, and A. Kornberg, *J. Biol. Chem.*, **174**, 979 (1948).
39. S. Korkes, A. del Campillo, and S. Ochoa, *J. Biol. Chem.*, **187**, 891 (1950).
40. J. N. Baptist, C. R. Shaw, and M. Mandel, *J. Bacteriol.*, **99**, 180 (1969).
41. J. N. Baptist, C. R. Shaw, and M. Mandel, *J. Bacteriol.*, **108**, 799 (1971).
42. K. C. Kendell and H. S. Goldberg, *Appl. Microbiol.*, **19**, 505 (1969).
43. R. A. D. Williams and S. A. Sadler, *J. Gen. Microbiol.*, **65**, 351 (1971).
44. R. Y. Stanier, "Toward an Evolutionary Taxonomy of the Bacteria," in A. Perez-Miravete and D. Pelaez, Eds., *Recent Advances in Microbiology*, Assoc. Mexicana Microbiologia, Mexico, D. F., 1971, p. 595.
45. R. A. Jensen and S. L. Stenmark, *J. Bacteriol.*, **101**, 763 (1970).
46. R. A. Jensen, *J. Bacteriol.*, **102**, 489 (1970).
47. H. B. LeJohn, *J. Biol. Chem.*, **246**, 2116 (1971).
48. H. C. Neimark, *Int. Ass. Microbiol. Soc. Abst.*, 13 (1973).
49. H. C. Neimark, *Annu. Meeting Amer. Soc. Microbiol. Abst.*, 108 (1974).
50. D. Bratthall, *Odontol. Rev.*, **21**, 143 (1970).
51. C. K. Williamson, "Serological Classification of Viridans Streptococci from the Respiratory

Tract of Man," in C. A. Leone, Ed., *Taxonomic Biochemistry and Serology,* Ronald Press, New York, 1964, p. 607.

52. A. L. Coykendall, *J. Bacteriol.,* **106,** 192 (1971).
53. J. L. King and T. H. Jukes, *Science,* **164,** 788 (1969).
54. J. Johnson, *J. Syst. Bacteriol.,* **23,** 308 (1973).
55. F. Cohn, *Beitr. Biol. Pflanz.,* **2,** 127 (1872).
56. F. Cohn, *Beitr. Biol. Pflanz.,* **3,** 141 (1875).
57. C. B. Van Niel, *Cold Spring Harbor Symp. Quant. Biol.,* **11,** 285 (1946).
58. C. B. Van Niel, "Classification and Taxonomy of the Blue-Green Algae," in California Academy of Science, San Francisco, 1955, p. 89.
59. *Bergey's Manual of Determinative Bacteriology,* 8th edit., R. E. Buchanan and N. E. Gibbons, Eds. Williams and Wilkins, Baltimore, 1974.
60. F. Gasser, M. Mandel, and M. Rogosa, *J. Gen. Microbiol.,* **62,** 219 (1970).
61. F. Gasser, *J. Gen. Microbiol.,* **62,** 223 (1970).
62. N. Sriranganthan, W. E. Sandine, and P. R. Elliker, *Annu. Meeting Amer. Soc. Microbiol. Abst.,* 30 (1974).
63. E. Margoliash and W. M. Fitch, "The Evolutionary Information Content of Protein Amino Acid Sequences," in Miami Winter Symposium 1, North Holland Publishing Co., Amsterdam, p. 33, 1970.
64. M. Goodman, J. Barnabas, G. Matsuda, and G. W. Moore, *Nature* **233,** 604 (1971).
65. N. Arnheim, Jr. and A. C. Wilson, *J. Biol. Chem.,* **242,** 3951 (1967).
66. N. Arnheim, Jr., E. Prager, and A. C. Wilson, *J. Biol. Chem.,* **244,** 2085 (1969).
67. M. F. Glaessner, *Biol. Rev.,* **37,** 467 (1962).
68. F. M. Swain, *Annu. Rev. Microbiol.,* **23,** 455 (1969).
69. E. M. Prager and A. C. Wilson, *J. Biol. Chem.,* **246,** 7010 (1971).
70. M. Reichlin, *Immunochemistry,* **11,** 21 (1974).
71. E. M. Prager, M. Fainaru, A. C. Wilson, and R. Arnon, *Immunochemistry,* **11,** 153 (1974).

CHAPTER 4

IMMUNOCHEMISTRY OF BACTERIAL ATPases

JOEL D. OPPENHEIM and MARTIN S. NACHBAR

1. INTRODUCTION

Bacterial membranes, like those of eukaryotic cells, are complex structures responsible for a wide variety of vital cell functions. Unlike eukaryotic cells, however, which possess a large number of membranous organelles each with a specialized function, bacteria contain only two major membrane structures, the cytoplasmic or plasma membrane and the mesosomal membrane. As Salton (1,2) has thoroughly discussed, of these two membrane systems, only the plasma membrane has been firmly established as a multi-component, multifunctional structure that is responsible for the performance of most of the cell functions. These functions include the transport of metabolites, electron transport and oxidative phosphorylation, and the biosynthesis of both comparatively simple molecules (e.g., exocellular enzymes and toxins) and complex macromolecular components of the bacterial capsule, envelope, wall, and membrane itself, such as phospholipids, lipopolysaccharides, and glycoproteins. In addition, the membranes are sites of the anchoring point and probably the replicating and separation point of DNA synthesis. Other specialized functions localized in membrane structures include photosynthesis, nitrification, and nitrogen fixation.

While "harboring" such a wide and varied range of functions within its structure, the bacterial membrane, as Salton (3) has pointed out, is constituted from classes of chemical components that are relatively few and simple; they are mainly comprised of proteins (both simple and conjugated) and lipids (primarily phospholipids and polar and apolar glycolipids), though some nucleic acids and polysaccharides have also been identified.

The manner by which these chemical components are associated is similar to those associations found in eukaryotic cells, a topic recently well reviewed by Nachbar et al. (4). The Singer and Nicolson (5) fluid mosiac model for membrane structure, while primarily based on information gained from the study of eukaryotic membranes, seems equally applicable to prokaryotic membranes.

For the investigator, the study of bacterial membranes offers two advantages over eukaryotic cell membranes. These advantages are the comparative ease of isolating the cytoplasmic and mesosomal membrane systems and the high yield of membrane preparations attainable. Despite these advantages, however, comparatively little is known of the structure-function relationships that exist in either membrane system, a situation that we believe has resulted partly from the so far meager use of immunologic techniques combined with biochemical and morphologic studies. In the following sections, we will first briefly discuss some of the basic techniques utilized in the study of biologic membranes on a chemical, physiologic, and structural basis. We will then focus our attention on the study of one particular membrane component, the membrane ATPase from *Micrococcus lysodeikticus,* which has been a valuable "model" component, extensively characterized at biochemical, functional, structural, and immunochemical levels. With the emphasis on immunolgic techniques, we will then illustrate how the use of this basic methodology in a multifaceted study of bacterial membrane ATPase can yield a great deal of information as to both the structural and the functional organization of membranes.

2. METHODOLOGY IN MEMBRANOLOGY

2.1 Biochemical Techniques

While a great deal of information has accumulated over the years concerning the lipid components and their organization in both eukaryotic and prokaryotic membranes, comparatively little has been gleaned about the types and numbers of the protein constituents. One technique that has been invaluable in identifying and characterizing membrane proteins has been polyacrylamide gel electrophoresis. In the application of this technique, membrane preparations are solubilized by treatment with the detergent sodium dodecyl sulfate, and during such solubilization, the released proteins are dissociated into their individual polypeptide subunit chains. They are subsequently separated in gels in the presence of the dissociating detergent according to their molecular weight due to negation of their charge and the molecular sieving property of the gel. Although such an analysis of a specific membrane preparation yields a characteristic pattern for the variety of protein subunits in that preparation, this method does not generally give information about the specific identity of the functions of any of the membrane components, whether enzymatic or structural. An alternate electrophoretic technique that can be used in the analysis of solubilized membrane components in the absence of detergent is also available. In such a system (Tris-glycine system), separation of undissociated proteins is based on both the total charge of the molecule in the particular buffer system used and on the molecular sieving action of the gel. One major advantage of this system is that, due to the absence of dissociating agents, enzymatic activity in biologically active proteins may still be retained after electrophoresis, and such activities can be measured either directly in the gel body by appropriate staining procedures or in assay on elution from the gel matrix. Proteins that are closely associated with lipids, however, may tend to aggregate in this system and may therefore be unresolved. At present, the primary applications of gel techniques are in monitoring cell membrane preparations during their isolation and in the establishment of purification and subunit nature of individual components.

The biochemical localization of most membrane-associated enzymes has been achieved primarily by the isolation of membrane fractions and the subsequent determination of enzymatic activities in the various fractions. (The purifications of specific enzymes are also based on this initial step.) Such studies are considerably more complex in most eukaryotic cells (red blood cells being the notable exception), where it is first necessary to distinguish between the various membranous organelles. In bacteria, however, with their much lower degree of intracellular differentiation, this problem is not so difficult to overcome. In all studies, membranes are first prepared from intact cells by disruption procedures that range from mechanical fragmentation by sonication, homogenization, or freezing and thawing to osmotic disruption by either direct osmotic lysis (e.g., red blood cells) or by indirect methods, such as in the preparation of osmotically sensitive forms of the cell, for example, protoplast or spheroplast formation of bacteria. After disruption, the various membrane fractions are concentrated primarily by differential centrifugation or gradient centrifugation techniques. Fractions are then assayed for enzymatic activity either directly or are pretreated with organic solvents or detergents to expose "buried" enzymes. By utilization of such techniques, a fairly clear picture of the distribution of enzymes as to their physical domains within isolated membrane fractions has been achieved in cells of both eukaryotic and prokaryotic origins. The interested reader is directed to the work and discussion by several investigators in this field, including Racker (6), on the localization of enzymes in liver mitochondrial membranes, Machtiger and Fox (7) for a complete review on the localization of enzymes and other proteins from bacterial membranes, and Nachbar and Salton (8) for the distribution of enzymes in *M. lysodeikticus* membranes. The critical evaluation of the results of such localizations, however, poses many problems that are difficult to ascertain, in that the basic assumption in such procedures is that the membrane-associated state of a specific enzyme reflects the natural *in situ* localization of the enzyme in the intact cell. This assumption obviously may not be valid, in that membrane components may be either gained or lost during the course of cell disruption, fractionation, and chemical treatment. Some proteins may be in equilibrium between a "soluble" and "membrane-bound" form, and this equilibrium could easily be shifted by changes in the aqueous environment; other proteins might be tightly bound through hydrophobic interactions and would be little affected by such treatment. The use of membrane-stabilizing agents, such as divalent cations, glutaraldehyde, or other cross-linking reagents, could result in associations that do not exist *in situ*. A thorough discussion of these and other problems can be found in two recent reviews by Nachbar and Salton (8) and Salton (9).

2.2   Immunologic Techniques

Classic immunologic techniques, while generally useful in the classification of bacteria, as discussed by Kwapinski (10), and more specifically in the analysis of the surface components of the bacterial cell, such as the capsular polysaccharides of gram-positive organisms, recently reviewed by Heidelberger (11), and the cell wall lipopolysaccharides of gram-negative bacteria, reviewed by Nikaido (12), have not been extensively utilized in studying bacterial membranes. Salton (13) has suggested the usefulness of such techniques (e.g., the use of specific antibody to membrane-specific antigens, such as enzymes) for following isolation and characterization and for the elucidation of membrane structures. One of the reasons for the limited use of such techniques is the less than overwhelming success that has been reported in the literature on bacterial and eukaryotic cell membranes so studied. The detection of a relatively small number of bacterial membrane antigens in solubilized preparations of two different organisms, by either immunodiffusion or immunoelectrophoresis testing against whole antimembrane sera, has been reported by Fukui et al. (14) and Kahane and Razin (15). The simplicity of these immunologic results strongly contrasts with the complexity of protein-staining patterns, as seen in the polyacrylamide gel electrophoresis of membrane preparations from these organisms by Salton et al. (16) and by Rottem and Razin (17), and with the expectancy that a large number of specific proteins would be found in a multifunctional structure, such as a bacterial membrane. Explanations for these discrepancies in relative sensitivity of these two different testing systems have been proposed by several investigators and include the possibility of poor immunogenic properties of membrane proteins due to their hydrophobic nature (15), the possession of common peptides or antigenic determinants among the various components (14), which has been established by Fukui and Salton (18) in analyses of three isolated membrane components, and the possible loss of antigenicity of the proteins due to denaturation or conformational changes caused during their solubilization, as suggested by Whiteside and Salton (19). Other explanations to be considered are that while bacterial membranes most certainly have many protein components, most of their antigenic determinants are buried within the membrane and are therefore either not available for immunologic processing in the antibody-producing host or are difficult to dissociate and to keep in solution during immunologic testing (14). Finally, the relative concentrations of many of the specific protein components in the membrane may be too low, so that antibodies are either not produced against them or, if produced, are in such low concentrations in the serum that they defy detection by standard means. It should be remembered that while double diffusion is one of the most commonly used immunologic techniques for the visualization of antigen-antibody reactions, it is dependent on the diffusional ability of the components to yield precipitates in a rather limited physical space in agar. Immunoelectrophoresis, while adding an electrophoretic separation step that greatly increases the resolution so that more complex mixtures of antigens can be analyzed, still does not totally overcome this problem, especially if one takes into account that many membrane components may have similar amino acid sequences and thus similar charge properties. New techniques that are readily available, such as crossed-immunoelectrophoresis (i.e., "jet" electrophoresis), which so far have not been widely tested, may further increase the resolution, so that more than only two or three membrane constituents can be identified.

One way of overcoming some of these obstacles would be the biochemical purification of individual components and the subsequent production of specific antibody against them. While this simplistic approach seems obvious, to date only a few bacterial membrane-associated enzymes have been purified to the homogeneity required to yield a specific homogeneous antiserum preparation. Such specific antibody could then be used to determine both the gross and fine distributions of its eliciting antigen either directly (e.g., agglutination or complement fixation) or by means of chemical (e.g., ^{125}I-labeled antibody) or visual amplification (e.g., fluorescence or ferritin-labeled antibody) of the immune reaction. Gross distribution may be detected in a fashion similar to procedures used for enzymatic localization, by use of neutralization of enzymatic activity rather than precipitation reactions as a criterion. As Cinader (20) has pointed out, most enzymes are at least partially inactivated by their specific antisera, indicating that such studies should be feasible. The distribution of any membrane component may also be determined through the use of radioactively labeled antibody. Day (21) has recently discussed quite extensively the use of such labeled antibodies for both *in vitro* and *in vivo* studies. Fine distribution studies can be performed via use of immunofluorescence and immunoelectron microscopy techniques and will be discussed in a later section of this review.

2.3 Electron Microscopy

Electron microscopy techniques have been extensively used in the study of biologic systems and have yielded a plethora of information on the general architectural arrangement of cellular structures for correlation with the major components of all cell types, including those of bacterial cells. In most bacteria, thin sections of intact cells reveal two distinct membranous domains, the limiting protoplast or plasma membrane and the mesosomes. They lack membranous organelles, such as mitochondria, Golgi, and endoplasmic reticulum. Some bacteria do contain specialized intracellular membranes that appear to be involved in specific physiologic functions. All of these membranes, however, show the familiar double-track unit membrane appearance in thin section and do not display any distinctive features that would enable the observer to differentiate between these isolated membrane fractions (notable exceptions being the purple membranes and the gas vacuoles of the obligate halophiles). Little more information can be obtained from negatively stained membrane preparations. Most mesosomal and specialized intracellular membranes, while exhibiting distinct structural morphology (i.e., tubular, flat, or vesicular), appear amorphous in nature (22, 23). Isolated bacterial plasma membranes examined by negative staining reveal the presence of a large number of particles on the membrane surface very similar to the stalked particles seen on mitochondrial membranes (24,25). Freeze-etching techniques applied to intact bacteria and isolated membrane fractions substantially verify the results seen in negatively stained preparations (see Fig. 8A), although the fracture faces of the internal cleavage planes also reveal a high degree of internal organization of globular structures (26–28).

While direct electron microscope observation of bacterial membranes has yielded valuable information about the comparative anatomy of such structures and has emphasized the basic similarities in a variety of cells, little can be deduced about the specific molecular architecture from the sole use of this technique. Electron microscopy must be combined with highly selective biochemical and immunologic techniques that can be used to identify specific membrane components. As Salton (1) has noted, such problems in localizing cellular membrane structures are not unique to bacteria, and the methods applicable to eukaryotic cells can also be used with prokaryotic membranes. One such approach has been the combination of cytochemical staining with electron microscopy. By the use of this combined approach, several specific enzymes in the membranes of bacteria have been localized with varying degrees of success. The localization of electron transport components, such as succinate dehydrogenase, has been reported by van Iterson and Leene (29), who concluded that such components were localized in the mesosomes on the basis of tetrazolium and tellurite staining by deposition of stain on membranes. In contrast, Sedar and Burde (30) found that the succinate-dependent tetrazolium staining occurred uniformly on both the mesosomal and plasma membranes. Voelz and Ortigoza (31), who employed lead salt precipitation staining, localized adenosine triphosphatase in the cytoplasmic membrane of *Myxococcus xanthus*. The peripheral localization of cell surface or periplasmic nucleases and phosphatases has similarly been achieved by Nisonson et al. (32) and Cheng et al. (33). In all of these studies, however, the results were less than definitive, and the resolution was rarely good enough to indicate whether the enzyme was on the outer or inner surface of the membrane, let alone to indicate the precise location and distribution of individual molecules.

2.4   Immunoelectron Microscopy

The use of immunologic techniques in conjunction with fluorescence microscopy or electron microscopy has provided still another means for elucidating membrane structure. The development by Coons and coworkers (34) of the immunofluorescence technique, in which antibody labeled with fluorescein is used as an immunologic tool to identify antigens on and within cells, opened the way for such studies at a light microscopy level. The covalent coupling of the electron-dense molecular ferritin to antibody without loss of immunologic activity by Singer (35) extended the microscopic identification and localization of antigens to the ultrastructural level. Numerous modifications and uses of the immunoferritin labeling technique have been developed over the last 15 years. The introduction by Hämmerling and coworkers (36) of the hybrid antibody technique, in which the divalent Fab$'_2$ antibody contains one combining site for a cellular antigen and one active site for ferritin, has opened the way for more specific labeling, in that one ferritin molecule corresponds to a single antigenic site. This method also avoids several of the undesirable consequences of coupling ferritin to antibody chemically and has permitted the use of visual markers other than ferritin (37, 38); this technique has also provided the potential for the visualization of at least two distinct antigens simultaneously, a feat recently reported by several investigators, including Lamm et al. (39), Neauport-Sautes et al. (40), and Wofsy et al. (41). The use of freeze-etching techniques by Karnovsky et al. (42), the use of direct labeling of intact ghosts on grids by Nicolson and Singer (43), the reconstruction of three-dimensional images from serial sections by Stackpole et al. (44), and the use of scanning electron microscopy of labeled whole cells, recently reported by Wofsy et al. (41), have further advanced the use of the immunoferritin techniques, so that topographic studies can be performed on large areas of the membrane rather than on the limited areas generally available in thin sections. Another interesting approach, the immunoenzyme technique, has been developed independently by Nakane and coworkers (45,46) and by Avrameas and Uriel (47). In this technique, antibody is coupled to a relatively stable enzyme, such as horseradish peroxidase, for which there exists a cytochemical detection method. For peroxidase, detection is based on the reaction of the enzyme with its substrate, hydrogen peroxide, and the subsequent reconstitution of the oxidized enzyme to the reduced form by an electron donor, such as diaminobenzidine. The free bonds of the oxidation product of diaminobenzidine rapidly react with each other to form an amorphous insoluble phenazine polymer that can be localized at the ultrastructural level. For the sake of simplicity, any procedure that utilizes electron-dense markers coupled to antibody for the localization of specific cellular molecules at the electron microscope level is referred to as an immunoelectron microscopy technique. The uses and techniques employed in immunoelectron microscopy have been the subject of numerous review articles in recent years by Andres et al. (48), Morgan (49), Wagner (50), and Sternberger (51) to which the interested reader can refer.

In all microscopic localization studies, whether by immunofluorescence or immunoelectron microscopy, the tagged antibodies are allowed to react with either whole cells or isolated fragments. The samples are then "processed" (i.e., immersed in an enzymatic reaction mixture, fixed, embedded, sectioned, and/or negatively stained, and/or subjected to freeze-etching techniques) and then observed under the microscope. Due to the much greater resolution of electron microscopy as compared to fluorescence light microscopy, the degree of correlation between labeled antibody and site of binding is much more precise and, at least in membrane architectural studies, makes electron microscopy the method of choice. Even with these refined techniques, however, such studies are based on the initial antigen-antibody interaction, and, as discussed earlier, adequate immunogenic membrane markers may be few in number or difficult to distinguish from one another by *in vitro* immunodiffusion tests because of close association in the form of membrane aggregates.

While a massive amount of information has accumulated in the literature on the immunoelectron microscope localization of antigenic determinants in mammalian tissues and cells (e.g., virus-induced host-specific and virus-specific surface antigens; red blood cell antigens, including ABO and Rh blood group antigens, complement components, viral receptor substances, IgG autoantibodies, and cell membrane components, such as torus protein, spectrin, and cholesterol; free mammalian cell antigens and associated molecules, including the H-2, θ, TL,HLA alloantigens, immunoglobulins, and microglobulins; and mammalian solid tissue antigens, including blood group antigens, insulin receptors, and undefined species- and tissue-specific antigens), comparatively few successful studies have been reported with bacterial components used as the antigenic source. Of the reports that have appeared, most have been concerned with bacterial cell wall and capsular components (see the previously cited review article by Wagner (50) for a complete listing of such studies), and only a few have dealt specifically with the localization of membrane antigens. By use of antibody directed against purified lipopolysaccharide somatic antigen of *Salmonella typhimurium,* Shands (52) reported the localization of this antigen on both sides of the cell wall and on the outer surface of the plasma membrane in osmotically shocked spheroplasts. While this was the first successful report of the localization of a bacterial membrane antigen, the quality of the micrographs made it difficult to judge the specificity of the labeling. In a similar study, Coulter and Mukherjee (53) reported the localization of α-toxin on cytoplasmic membranes within disrupted staphylococcal cells, but, again, the resolution was insufficient to firmly establish specificity. Nakane and coworkers (54) have been able to localize to a limited extent the amino acid transport proteins in the membrane envelope of *E. coli* through the use of fluorescent and peroxidase-labeled antibody conjugates. It is difficult to deduce from this study whether the binding is confined to the inner and/or outer membranes of the envelopes or to the periplasmic space. An interesting approach for the localization of cholesterol in both intact and fractionated membranes from erythrocytes and some mycoplasmas has been described by Pendleton et al. (55). In this study, membranes were first treated with a bacterial hemolysin from *Bacillus cereus,* cereolysin, which specifically binds to cholesterol, and were then treated with ferritin-conjugated antitetanolysin (an antiserum to a lytic toxin from *Clostridium tetani* that is antigenically very similar to cereolysin). A uniform distribution of ferritin was observed on the membranes of *Mycoplasma gallisepticum,* which contain relatively large amounts of cholesterol, whereas no labeling was observed in bacterial membranes that lack cholesterol. As the investigators have noted, the method has limitations, and some membranes in which cholesterol was not detectable were nonspecifically stained. To date, only one study, that by Oppenheim and Salton (23), has clearly established the localization and distribution of a specific bacterial membrane protein, the membrane-bound form of adenosine triphosphatase (ATPase), by use of ferritin-labeled antibody. This study will be fully discussed in a later section of this review.

3. BACTERIAL MEMBRANE ADENOSINE TRIPHOSPHATASE

Of all the membrane-associated enzymes, ATPase has been one of the most intensely studied in cells of both eukaryotic and prokaryotic origin. ATPases of differing metal ion requirements have been shown to be ubiquitous and important components of biologic membranes of such diverse origins as red blood cells (56, 57), the inner mitochondrial membrane (58,59), and the plasma membranes of bacterial cells (60–63). The Na^+-K^+ ATPases, which are distinct from mitochondrial and bacterial Mg^{2+}-Ca^{2+} ATPases, were the topic of a recently published symposium conducted by The New York Academy of Sciences (64).

3.1   Purification

As Salton (9) has noted, the bacterial membrane-associated ATPases have been so widely studied primarily because they have proved to be the easiest enzymes to be released from the membranes and the most amenable to purification techniques. The investigations of bacterial membrane ATPases were made possible by the development of procedures introduced by Weibull (65) for the isolation of membranes following the dissolution of the cell wall and the subsequent lysis of the osmotically sensitive "protoplasts." Subsequently, it became apparent from studies with a variety of bacterial membranes that ATPases can be released from their membrane-associated state by relatively gentle perturbations of the membranes. Ishikawa and Lehninger (66) were the first to note that when sonicated fragments of *M. lysodeikticus* membranes were suspended in distilled water, a soluble protein fraction was released that contained ATPases and other factors. Following the solubilization of ATPase by this water "shock" technique, Abrams (60) developed a more highly selective release procedure for the purification of ATPase from the membranes of *Streptococcus faecalis,* based on the cation depletion of membranes by washing in the absence of Mg^{2+}. Muñoz et al. (63,67) developed a further modification of the shocking procedure that effected the release of the bulk of the ATPase activity from membranes of *M. lysodeikticus*. In this procedure, the isolated membranes were washed in 0.03 M Tris hydrochloride buffer in the absence of Mg^{2+} prior to shocking in buffer of low ionic strength (0.003 M Tris). The combination of cation depletion and osmotic shocking of membranes has provided the basis for the selective release of ATPases from several other bacterial membrane systems in both gram-positive (68–70) and gram-negative bacteria (71–77). Alternative means for releasing ATPase from bacterial membranes are the sodium dodecyl sulfate (or sodium lauryl sulfate) technique for dissociation of membranes by Weinbaum and Markman (78) and Evans (79), Triton X-100 extraction, as described by Hanson and Kennedy (80), and organic solvent extraction, as used by Salton and Schor (81,82). The selective low-ionic-strength shock-wash procedure, however, has advantages over these two procedures, in that it avoids the presence of potentially deleterious agents. After release from the membranes and subsequent separation of the residual membrane, ATPases can be purified by fairly conventional biochemical methodology, which utilizes precipitation, differential and gradient centrifugation, adsorption and release from ionic columns (e.g., diethyaminoethyl cellulose), and molecular sieve chromatography. These techniques and additional ones have been fully discussed by Salton (9). By use of such techniques, relatively large quantities (>25 mg) of highly purified ATPase (> 100-fold purification) have been prepared from *M. lysodeikticus* membranes by Oppenheim and Salton (23).

The homogeneity of ATPases purified from bacterial membranes has been determined by various criteria generally used for enzymes and other proteins, which include the detection of a single sedimenting peak by ultracentrifugation analysis, the presence of a single or major band of protein and/or specific ATPase activity on electrophoresis in standard polyacrylamide gel systems, uniformity of particles as seen in the electron microscope, and homogeneity on reaction with specific antibody in agar double-diffusion and immunoelectrophoresis testing (see Fig. 1). By use of one or more of these criteria, membrane ATPase preparations from several different organisms have been judged as being homogeneous.

3.2   Physical Characterization

Purified bacterial ATPases have molecular weights that range from 300,000 to 400,000 daltons and have sedimentation coefficients between 12S and 15S (9). From sodium dodecyl sulfate (SDS) polyacrylamide gel electrophoresis data, reduced and alkylated purified bacterial ATPases appear to consist of primarily two major polypeptide chain subunits, termed α and β, of very similar or slightly different molecular weights, each in the range of 60,000–70,000 (except for the ATPase of *S. faecalis,* which has two subunits of 33,000 each) and present in a 1:1 ratio. In addition to the two major polypeptides, as many as three minor polypeptide bands have been seen in gels. The presence of these minor bands seems to be dependent on the mode of enzyme release and

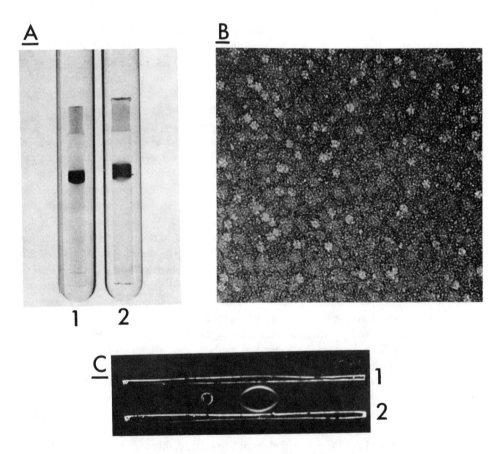

Fig. 1. Criteria of *M. lysodeikticus* ATPase homogeneity. A: Polyacrylamide gel electrophoresis of purified ATPase stained with Coomassie blue (A-1) and of ATPase enzymatic activity with lead acetate (A-2). Purified ATPase preparation negatively stained with ammonium molybdate as seen in the electron microscope. C: Immunoelectrophoresis of purified ATPase (center well) developed against anti-ATPase (trough 1) and antisonicated membrane sera (trough 2) From Oppenheim and Salton (23) and Whiteside and Salton (19). By permission of *Biochimica et Biophysica Acta* and *Biochemistry,* respectively.

purification. Salton and Schor (81) have shown that the ATPase released from *M. lysodeikticus* membranes by the *n*-butanol extraction procedure possessed only the α- and β-subunit bands, whereas the ATPase obtained by the shock-wash release method from the same membranes yielded a product that contained, in addition to the α and β subunits, minor proteins associated with it. Nelson and coworkers (75) have recently reported that when purified ATPase from *E. coli,* which exhibited, in addition to the α and β subunits, two minor polypeptides (γ and ε) in SDS gels, was treated with trypsin, followed by chromatography, it became an active ATPase preparation, consisting exclusively of α and β subunits. It is interesting to note that Futai et al. (76), who used a purification procedure only slightly different from that described by Nelson et al. (75), have reported the purification of an *E. coli* ATPase that contained one additional minor subunit (δ). The presence of this polypeptide produced an active ATPase enzyme that could both bind to deficient membrane and reconstitute ATP-driven transhydrogenase activity.

Schor et al. (83) have shown that both the *n*-butanol-extracted and shock-wash-released ATPase from *M. lysodeikticus* membranes have very similar amino acid compositions. This finding is not surprising when one considers that the major subunits (α and β) comprise the bulk of the polypeptides in one instance (shock-wash preparation) and the entire polypeptide composition in the other preparation (*n*-butanol extraction). The principal minor band seen in SDS gels of the shock-wash ATPase corresponds to the membrane component identified and purified by Fukui et al. (14), which migrated rapidly on electrophoresis in polyacrylamide gel and, because of its undetermined function, was simply referred to as "fast-moving component," or FMC. The subunit structure of the bacterial ATPase molecules has been further elucidated by electron micrographs of negatively stained enzyme preparations. Muñoz *et al.* (84) were the first to show that ATPase activity of *M. lysodeikticus* was associated with particles of approximately 100 Å diameter, which consisted of six peripheral subunits that surrounded a central structure, as can be seen in Fig. 1B. Subsequent studies by Ishida and Mizushima (85) and Schnebli et al. (86) on ATPases isolated from *Bacillus megaterium* and *S. faecalis* have confirmed these observations. From molecular weight determinations, subunit structure analysis in polyacrylamide gels, and appearance of the enzyme particles in the electron microscope, Schnebli et al. (86) proposed a model structure of ATPase from *S. faecalis.* In further studies on the *S. faecalis* ATPase system, Baron and Abrams (87) isolated a protein from the plasma membranes, called nectin, which was required for the binding of the enzyme to the membrane. Salton and Schor (81) have suggested that the fast-moving component from *M. lysodeikticus* membranes serves a similar noncatalytic, anchoring function in these membranes. Futai et al. (76) have subsequently identified a minor polypeptide component of *E. coli* ATPase (δ subunit) that corresponds to nectin and FMC and whose presence is required for binding to membranes deficient in ATPase and for ATP-driven transhydrogenase activity. The amino acid compositions of ATPases from different bacteria have been summarized by Salton (9). These data do not reveal any unusual features about their amino acid compositions and, in fact, show remarkable similarity.

3.3 Enzymatic Characterization

The ability to readily purify bacterial ATPases has subsequently led to their thorough biochemical characterization. Such characterization was a major topic in a recent review by Salton (9) and therefore will only be discussed briefly. The substrate specificity of all bacterial ATPases, by definition, is for ATP, although appreciable hydrolysis of GTP and, to a lesser extent, of CTP, UTP, TTP, and ITP have been observed. Each bacterial ATPase system, depending on the species, appears to have its own metal ion requirements, although all that have been characterized are Mg^{2+}- and/or Ca^{2+}-activated enzymes. In both membrane-bound and soluble forms, maximal activities are observed with these cations; they thus appear to resemble the mitochondrial ATPase rather than the ion transport ATPases, which respond to $Na^+ + K^+$ stimulation and are sensitive to ouabain. In addition to their role in activation, divalent cations appear to be involved in the attachment of the enzyme to the membrane; Abrams (60) and Ishida and Mizushima (85) have shown that their presence is an essential requirement for the reassociation of the soluble ATPase with the enzyme-depleted membrane. Bacterial ATPases are not inhibited by either ouabain or oligomycin at concentrations that inhibit the enzyme from other sources; they are, however, sensitive to the inhibitory effects of carbodiimides and to other common inhibitors, such as azide, cyanide, dinitrophenol, and sulfhydryl-blocking reagents. Optimal activity of membrane-associated and purified ATPase has been found in a broad range from pH 7 to 9. One interesting feature of ATPases is their marked stimulation of activity-upon treatment with trypsin. Masked or latent Mg^{2+}- and Ca^{2+}-activated ATPases in mitochondrial and chloroplast preparations have been observed by Racker (88) after such treatment. Trypsin stimulation of bacterial membrane-bound ATPase has similarly been observed in M. lysodeikticus, and, in fact, in its membrane-associated state, the ATPase molecule generally shows a relatively low level of enzymatic activity, unless activated by such proteolytic treatment, as has been shown by Muñoz et al. (63). The mechanism involved in this masking of hydrolytic activity of ATPase has not yet been determined. Trypsin stimulation of solubilized ATPase can also be observed and appears to be dependent on the mode of release of the ATPase complex and its purification. Salton and Schor (81) noted that ATPase obtained from M. lysodeikticus by the shock-wash procedure was stimulated by trypsin and also possessed rebinding capabilities, whereas ATPase released by organic solvent extraction procedures and purified was unable to rebind to depleted membranes. It was also noted in these studies that in SDS-polyacrylamide gel electrophoresis, shock-washed ATPase showed the presence of one or more minor protein bands in addition to the major α and β subunit bands, whereas the ATPase released by the organic solvent extraction procedure exhibited only the α and β subunit bands. As discussed earlier, Futai et al. (76) have reported similar results in studies on E. coli ATPase. Bacterial ATPases are generally stable when associated with membrane structures but when released and purified in soluble form, many exhibit cold lability. Evans (62) and Salton (9) have observed that the ATPases from E. coli and M. lysodeikticus in either membrane-associated or soluble form exhibit enhanced activities after pretreatment at temperatures between 40 and 60°C, although at temperatures above 60°C, the ATPases are rapidly inactivated, as are most other enzymes.

3.4   Function

The function of bacterial ATPases has not yet been firmly established, but from the number of similarities to mitochondrial and chloroplast ATPases, there are strong indications that these enzymes play an important role in oxidative phosphorylation. Harold (89) has suggested that they may serve to couple oxidative phosphorylation to ATP metabolism and also may be useful in the utilization of ATP as an energy source for a variety of membrane functions. Harold and Papineau (90) have also presented evidence that suggests that membrane-associated ATPase may serve to couple ATP metabolism to cation transport, in that ATPase from *S. faecalis* participates in proton extrusion. More recent studies with uncoupled *E. coli* mutants (*unc⁻*) by Butlin et al. (91) and Cox and Gibson (92) implicate the ATPases in processes of energy transduction in the membranes.

4. IMMUNOCHEMISTRY OF BACTERIAL ATPase

4.1 Background

As mentioned earlier in this review, Salton (13) has indicated the usefulness of an immunologic approach for the study of bacterial membranes. Immunologically detectable membrane-specific antigens could be used in tracing the fate of membrane components during isolation and fractionation and could also be of great value in the elucidation of the molecular organization and architectural arrangement in such membranes. By use of antisera prepared against cytoplasmic and membrane preparations from *M. lysodeikticus,* Salton (13) observed in double-diffusion precipitin reactions in agar that unwashed membrane fractions were contaminated with cytoplasmic antigens and that even by washing such membranes six times, which reduced cytoplasmic contamination, only a few membrane-associated antigens could be detected. Salton (13) subsequently utilized the membrane antisera to detect the release of these membrane-associated antigens after solubilization of membranes with detergents and an organic solvent. These results were confirmed and expanded by Fukui and coworkers (14). They observed that, by either the double-diffusion agar method or by immunoelectrophoresis, at the most there were three antigen components detectable from carefully washed membrane preparations from *M. lysodeikticus* and that the patterns of membrane antigens obtained on immunoelectrophoresis were not quantitatively different for membrane preparations dissociated by sonic oscillation at pH 9.0, treatment with 0.3% SDS or 0.3% Triton X-100, or by digestion with trypsin, phospholipase A, or phospholipase C. One of the major membrane antigens detected in these studies has been identified by Muñoz et al. (63) as a Ca^{2+}-dependent ATPase. Fukui and coworkers (93) subsequently isolated and established that the other two major antigenic components from *M. lysodeikticus* membranes were an NADH dehydrogenase and the fast-moving component (FMC) discussed earlier. Immunochemical analysis in agar diffusion tests of ATPase (purified by the butanol extraction procedure), NADH dehydrogenase, and FMC clearly demonstrated that each protein possessed an individual antigenic specificity when reacted against antisera to purified membranes. However, prior treatment of the proteins with 0.3% SDS revealed the presence of a common antigenic component in each. Moreover, it was also observed that trypsin digestion of each of the purified proteins resulted in peptide "fingerprints" that revealed the presence of similar major peptides, which were absent in the fingerprint of a purified cytoplasmic enzyme (catalase), thus indicating that at least certain membrane proteins may possess very similar peptide regions or chains and/or antigenic determinants. Similar results have been reported by Yang and Criddle (94) for three purified yeast mitochondrial membrane proteins. In further studies that employed tryptic fingerprint analysis, Fukui and Salton (18) found that 92% of the FMC protein could be accounted for by the common peptides and that ATPase was composed of 53% of such peptides. It is interesting to note that the amino acid composition of the purified FMC revealed that 42% of the amino acid residues were of a hydrophobic variety (18). The presence of a hydrophobic peptide sequence could provide a mechanism for ensuring attachment of such a protein to the membrane.

Another interesting aspect of the results reported by Fukui et al. (14) was the detection of a surface antigen on the outer membrane face of *M. lysodeikticus* that was shown by immunochemical analysis to contain a detectable common antigenic determinant with SDS-treated ATPase but not the intact untreated ATPase molecule. This phenomenon was observed by comparing the reactions of antimembrane serum, which was absorbed by prior treatment with protoplasts, and unabsorbed antiserum upon subsequent reaction against purified membrane proteins. It was shown that whereas antibodies to ATPase and FMC were not removed by absorption with protoplasts, in that after such absorption, unbound antisera could still react in immunodiffusion gels with either free ATPase, FMC, or solubilized membranes prepared from such protoplasts, the antigenic specificity demonstrable with SDS-treated ATPase disappeared upon such treatment. These results strongly suggest that the antigenic determinants exhibited by purified ATPase and FMC are either not localized on the outer surface of the membrane or are inaccessible. It would be reasonable to expect for functional purposes that the ATPase would be located on the inner surface. The internal localization of the intact ATPase molecule has been further supported by the studies of Oppenheim and Salton (23) by use of immunoelectron microscopy and by Salton et al. (95) with ^{125}I labeling. The fact that antigenic specificity was exhibited after treatment of the purified ATPase with SDS on the outside of intact protoplasts is an intriguing observation. Such a result could be related to the allotopic character of this membrane enzyme or may be explained by the presence of a masked antigen in the intact ATPase molecule, which is found in the unmasked form on the outer membrane surface, that becomes recognizable during the immunologic processing for antibody formation. Further evidence for the presence of a possible ATPase antigenic determinant on the outer membrane surface has been put forth by Monteil and associates (96, 97). These investigators have reported that antiserum prepared against purified ATPase from *Proteus* P18 L-forms induced rabbit antibodies that could both inhibit L-form growth on solid media and inhibit the enzymatic activity of membrane-bound or solubilized ATPase in enzymatic assays. Antisera prepared against *Proteus*, *Proteus* L-forms, *Proteus* L-form-shocked membranes, and crude enzyme, however, all showed greater growth inhibition than did the antiserum prepared against the purified ATPase. Antisera to the *Proteus* and *Proteus* L-forms had minor ATPase inhibition capacity, whereas the anti-*Proteus* L-form-shocked membrane and anticrude enzyme sera could, in fact, inhibit ATPase enzymatic activity to an even greater degree than could the antipurified ATPase serum. The results of these studies, while interesting, are somewhat clouded by the fact that the investigators did not exclude the possibility of the presence of an antibody species that might cross react between the purified ATPase and the outer membrane surface of the organism.

4.2 Antibody Production

After the purification of the Ca^{2+}-dependent ATPase from the membranes of *M. lysodeikticus* by Muñoz et al. (63,67), and its subsequent identification as a major membrane antigen, it became possible to prepare specific antisera against this enzyme. By use of highly purified ATPase that was judged homogeneous on the basis of polyacrylamide gel electrophoresis, electron microscope examination, and agar gel diffusion against antimembrane antisera (see Fig. 1), Whiteside and Salton (19) and Oppenheim and Salton (23) successfully prepared ATPase antisera in rabbits and guinea pigs by employing standard immunization techniques. Subsequently, Hanson and Kennedy (80) and Nelson et al. (75) prepared ATPase antisera to the purified *E. coli* enzyme and to the trypsin-treated product, respectively. Recently, Monteil and Roussel (97) have reported the production of antiserum to ATPase from *Proteus* P18.

4.3   Precipitation Analysis of ATPase-antibody Complexes

By use of the method of Mancini et al. (98), Whiteside and Salton (19) developed a radial immunodiffusion assay to quantitate the ATPase antigen. Figure 2 shows the results of such an assay when increasing amounts of purified enzyme were allowed to

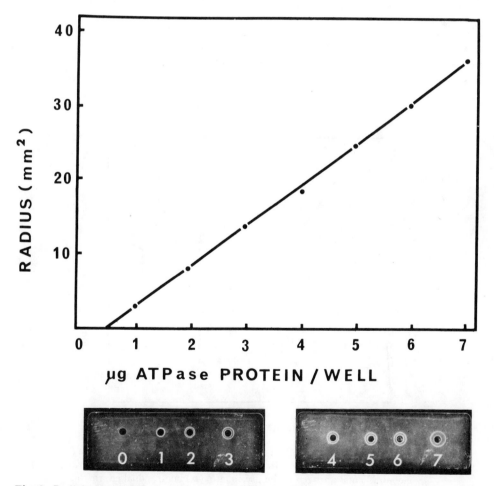

Fig. 2. Radial immunodiffusion slide assay of purified ATPase. From Whiteside and Salton (19). By permission of *Biochemistry*.

diffuse into agar gel that contained a known quantity of the specific anti-ATPase serum. The areas of the precipitin rings, determined from the diameters of rings by an ocular micrometer, were found to be proportional to the amounts of enzyme placed in the wells and, when plotted in terms of the radius squared versus micrograms of enzyme protein per well, showed a linear relationship. In addition to measuring the antibody-antigen reaction in gels, Whiteside and Salton (19) also performed a quantitative precipitation assay. A fairly characteristic precipitin curve was obtained, as can be seen in Fig. 3, when increasing amounts of purified ATPase were added to a constant amount of anti-ATPase γ-globulin. By monitoring enzymatic activity in the supernatants of each fraction after removal of the immune precipitates by centrifugation, it was shown that at the equivalence point, no ATPase activity was detectable, indicating that all of the enzyme had been precipitated, whereas enzymatic activity was found in those fractions in the antigen excess zone, as would be expected.

4.4 Inhibition of Enzymatic Activity by Antibody

4.4.1 Purified Solubilized ATPase

Most enzymes appear to be at least partially inactivated by their specific antisera, and, as Cinader (20) has noted, antigen-antibody reactions of enzyme systems can therefore be used as a quantitative tool in measuring such reactions, in conjunction with the more classic precipitation studies. In addition to the quantitative information obtained from such enzyme-antibody reactions, a great deal can also be learned about the nature of the

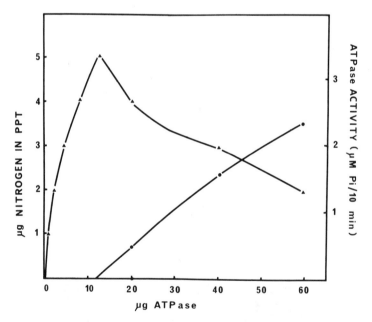

Fig. 3. ATPase anti-ATPase precipitin curve obtained by adding increasing amounts of purified enzyme to a constant amount of antisera γ-globulin. Total nitrogen in the precipitates (▲) and residual enzymatic activities (●) in each fraction were monitored. From Whiteside and Salton (19). By permission of *Biochemistry*.

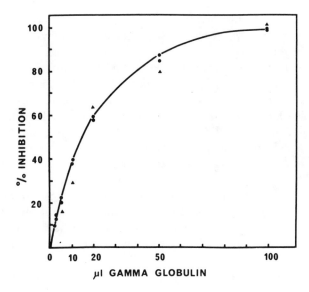

Fig. 4. Inhibition of soluble ATPase enzymatic activity by
its specific antiserum, showing that complete (100%) enzymatic inhibition can be obtained in the presence of excess
antibody. From Whiteside and Salton (19). By permission
of *Biochemistry*.

enzyme itself. As Cinader has noted (20,99), by studying the interactions between
enzyme, specific antibody, and catalytic substrate, the mode of inhibition can be
determined. Enzymatic inhibition may be due to a physical steric blockage of the
enzyme catalytic site that would prevent entrance of substrate. Such blockage may be
due to either the direct interaction of antibody with an antigenic determinant within the
catalytic site or, indirectly, to antibody-antigen complexes formed at an adjacent site,
which would overlap the catalytic site. Alternatively, inhibition could be caused by a
conformational change at the active site that resulted from an alteration in the enzyme
molecule due to the interaction with its specific antibody, in which case substrate would
no longer be accepted. The mechanism of inhibition in most systems studied appears to
support the concept of direct or indirect steric hindrance of the binding of substrate at
the catalytic site rather than that of conformational changes in the enzyme molecule.
Both Cinader (20) and Arnon (100 have shown, by use of different enzyme systems, that
enzymatic inhibition is at least partially dependent on the molecular size of the substrate
molecule, in that complete or nearly complete inhibition is observed with a
macromolecular substrate, whereas a smaller substrate molecule yields significantly less
inhibition. Thus, binding of antibody near, but not necessarily at, the enzyme catalytic
site would exclude a large-molecular-weight substrate, while a smaller molecule could
still gain access.

   The ability of specific anti-ATPase serum to inhibit the liberation of P_i from ATP by
solubilized bacterial ATPase was initially observed by Whiteside and Salton (19) with
antiserum prepared against ATPase purified from *M. lysodeikticus*. The inhibition of
soluble ATPase activity by specific antiserum in the *M. lysodeikticus* system is shown in
Fig. 4. As can be seen, complete inhibition of enzymatic activity is observed at a suitably
high concentration of anti-ATPase γ-globulin. This phenomenon of complete inhibition

of enzymatic activity by reaction with an excess of its specific antibody, though typical of low-molecule-weight hydrolytic enzymes (e.g., ribonucleases, lysozymes) that act on high-molecular-weight substrates, is somewhat unusual for such a large enzyme as ATPase with a relatively low-molecular-weight substrate. Similar results, however, have been reported for the effect of specific antisera on ATPases from other bacteria [*E. coli* (75,80) and *Proteus* (97)] and from those of eukaryotic origin [beef heart and yeast mitochondria, spinach chloroplasts (10)], thus indicating a possible common mechanism of enzymatic inhibition for all ATPases. To determine the mechanism of enzyme inhibition, Whiteside and Salton (19) performed a series of competitive inhibition experiments, the results of which can be seen in Fig. 5. In both preincubation and simultaneous addition experiments, the kinetics of inhibition indicate that the anti-ATPase acted as a noncompetitive inhibitor with respect to the enzyme substrate, in that the degree of inhibition caused by a given amount of the antibody was independent of substrate concentration. Such an inhibitory response would indicate an indirect physical exclusion of substrate to the catalytic sites rather than a direct blockage of the

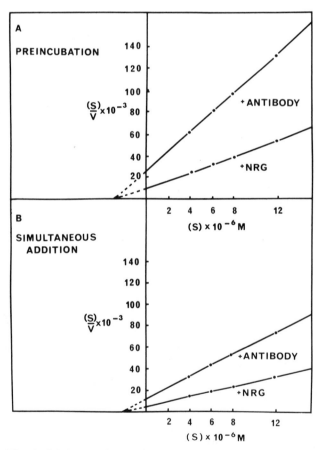

Fig. 5. Inhibition of soluble ATPase enzymatic activity by its specific antiserum in the presence of enzyme substrate (ATP) in preincubation (A) and simultaneous addition (B) experiments that show kinetics of noncompetitive inhibition. From Whiteside and Salton (19). By permission of *Biochemistry*.

active centers of the enzyme by antibody molecules. It is interesting to note that in the simultaneous addition experiment, the anti-ATPase appeared to be a less effective inhibitor when combining with ATPase in the presence of the substrate. Cinader (20) has noted such substrate protection in several enzyme-antienzyme systems. In the only other inhibition kinetics study that employed a bacterial ATPase-antibody system so far reported, Monteil and Roussel (97) observed an uncompetitive type of inhibition. In their *Proteus* ATPase-anti-ATPase system, antibodies do not appear to be blocking the catalytic sites but, instead, are interfering with the enzymatic reaction by causing a conformational change in the enzyme molecule that alters the active site. These investigators also did not observe the phenomenon of substrate protection.

In another interesting approach for the use of anti-ATPase serum, Hanson and Kennedy (80) used antiserum prepared against solubilized purified *E. coli* ATPase to test the hypothesis that this enzyme is coupled to the energy-linked transhydrogenase of that organism. In inhibition experiments, they observed that the stimulation by ATP of the transhydrogenase activity is correspondingly diminished as ATPase is inhibited by increasing antibody concentration. Nelson et al. (75) have also shown that antisera prepared against purified α and β subunits isolated after trypsin treatment of solubilized Mg^{2+}-Ca^{2+} ATPase from *E. coli,* at suitable concentrations, inhibited both the native enzyme and the trypsin-treated enzyme and also inhibited the parallel ATP-driven transhydrogenase reaction. The immunologic data, along with results observed on the kinetics of chemical inhibition of ATPase (75,80) and the isolation of *E. coli* ATPase mutants (102–104) that have greatly reduced levels of membrane ATPase and transhydrogenase activity, strongly suggest that these two enzymatic reactions are, in fact, coupled, even though they are catalyzed by different enzymes.

4.4.2 Membrane-bound ATPase

The ability of anti-ATPase serum prepared against the purified solubilized enzyme to inhibit the membrane-bound form has been shown by several investigators (23,75,80,96). As can be seen in Fig. 6, at least 95% of the membrane-bound ATPase enzymatic activity on *M. lysodeikticus* membranes, as determined by trypsin activation assays, could be inhibited with sufficiently high concentrations of specific antiserum prepared against solubilized ATPase (23). Similar results have been reported for the *E. coli* ATPase-anti-ATPase system (75,80). The inability to obtain 100% enzymatic inhibition of the membrane-bound form of the enzyme in the *M. lysodeikticus* system, as was observed with the soluble ATPase (19), could be readily explained by the nature of the enzymatic assay needed to detect the enzyme on the membrane when preincubation with trypsin is required. This proteolytic enzyme could conceivably react with some of the antibody molecules that had initially blocked the ATPase catalytic site (as previously discussed), thus allowing a fraction of the enzymatic reaction to occur. It is also interesting to note that in the *M. lysodeikticus* membrane-bound enzyme inhibition assay, assuming that membrane-bound ATPase comprises approximately 10% of the total protein content of thoroughly washed membranes, as Muñoz et al. (63) have calculated, a considerably larger quantity of the specific γ-globulin fraction of the antisera was required to achieve inhibition (23), as compared to the results found with the solubilized enzyme (19). Hanson and Kennedy (80) have alluded to a similar finding in the *E. coli* system. One obvious explanation of these observations may be that the anti-ATPase sera could be reacting with membrane components, other than ATPase, which in the *M.*

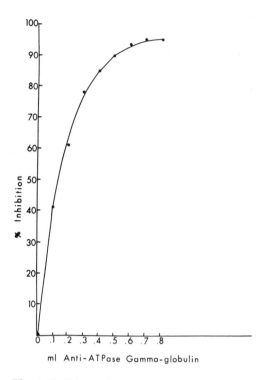

Fig. 6. Inhibition of membrane-bound ATPase by specific antisera to soluble ATPase. Complete enzymatic inhibition could not be obtained, even with a large antibody excess. From Oppenheim and Salton (23). By permission of *Biochimica et Biophysica Acta.*

lysodeikticus system have been shown by Fukui et al. (14), under certain conditions, to include common antigenic determinants. This explanation appears highly unlikely, however, in that when antisera were reacted against solubilized membranes (SDS-solubilized *M. lysodeikticus* membranes and Triton X-100 *E. coli* membranes) and purified ATPase preparations, only single precipitin lines for each preparation, which showed reactions of identity, were observed in gel double-diffusion plates in both the *M. lysodeikticus* (23) and *E. coli* (80) systems, respectively. A more plausible explanation may be that because the anti-ATPase sera used in these assays were prepared from solubilized ATPase, several of the antigenic determinant sites located on the molecule may no longer be available to react immunologically when the enzyme is positioned on the membrane in its bound form, thereby minimizing the chances of an enzymatic neutralization unless a large excess of antisera in relation to membrane protein was added.

4.5   Effects of ATPase Dissociation on Reaction with Antisera

Dissociating agents, such as SDS, urea, and guanidine hydrochloride, have been widely employed in the study of subunit structure of purified enzymes. The most common usage of such agents is in polyacrylamide gel electrophoresis, as discussed earlier. These agents, however, can also be used for studying the enzymatic activity and antigenicity of the enzyme molecule by direct assay procedures. As Rottem and Razin (105) have observed in membrane proteins isolated from *Mycoplasma,* dissociation into subunits can result in a loss of enzymatic activity and/or antigenicity of such treated molecules. Muñoz et al. (63) reported that dissociation of the solubilized ATPase molecule from *M. lysodeikticus* into 3.5S subunits occurred on treatment with 1% SDS. Whiteside and Salton (19) subsequently determined that in the dissociated state, ATPase is not only enzymatically inactive but is also incapable of reacting with anti-ATPase or with antisera to whole untreated membranes in agar gel, even after prolonged incubation. Fukui et al. (14), however, observed that such detergent-dissociated ATPase could exhibit a precipitin line with the antiserum to sonicated membranes and appeared to possess a common antigenic determinant with an unidentified outer surface membrane antigen. It was further noted that neither enzymatic activity nor the ability to react with anti-ATPase was restored by the removal of the SDS by dialysis (19).

Dissociation of the ATPase with 2.6 M guanidine hydrochloride yielded products similar to those obtained with SDS with respect to migration in acrylamide gel, response to specific anti-ATPase and antimembrane antisera, and loss of enzymatic activity (19). It has thus been suggested by Whiteside and Salton (19) and Fukui et al. (14) that a masked antigenic site may exist in the intact ATPase molecule that can be uncovered by sonication or dissociation with 1% SDS or 2.6 M guanidine hydrochloride. It is interesting to note, however, that treatment of ATPase with 1.3 M guanidine hydrochloride, while resulting in a dissociation product that migrated faster in polyacrylamide gel electrophoresis than did untreated ATPase, caused no loss in antigenicity against anti-ATPase serum and only a 60% reduction in enzymatic activity. This dissociation was reversible, in that the removal of the guanidine hydrochloride by dialysis restored both the enzymatic activity and the electrophoretic mobility typical for the untreated enzyme (19).

4.6   Use of Antibody to ATPase in Taxonomic Studies

Until recently, most comparative bacterial serology relied primarily on the antigenic characteristics of cell surface components, such as capsules, cell walls, and flagella (10,11). Relatively few specific bacterial membrane proteins, especially those characterized as enzymes, have been studied from a comparative immunologic basis, because such components were more difficult to isolate and purify. Once such proteins were purified and characterized, the antisera prepared against them offered invaluable diagnostic reagents that could be used for taxonomic classification. Argaman and Razin (106) successfully utilized such an approach to study the antigenic properties of the Mycoplasmas and have shown that such studies are of diagnostic value for this group of organisms. The elegant work of Stanier and coworkers (107) on the immunologic properties of two inducible enzymes in the β-ketoadipate pathway of *Pseudomonas* sp. have clearly shown that proteins can also be used as antigens for serotaxonomic studies. As discussed by Wilson and Kaplan (108), the immunologic comparison of enzymes offers many advantages in taxonomic studies, in that besides being able to use classic qualitative precipitin reactions in gels to detect cross reactions among homologous enzymes from other species, several quantitative methods are also available, including enzyme inhibition and microcomplement fixation, where the degree of such cross reactivity or immunologic homogeneity can be precisely determined.

Whiteside and Salton (19) first reported the use of antisera specific for the ATPase from *M. lysodeikticus* in taxonomic studies. This work was subsequently expanded in a second publication by Whiteside et al. (109). Figure 7 shows some of the reactions in agar gel of *M. lysodeikticus* anti-ATPase serum against membrane ATPases isolated by the selective release method from other pigmented micrococci and sarcinae. Different degrees of cross reaction were observed among the various bacterial ATPases. *Micrococcus tetragenus, Sarcina flava,* and *Sarcina lutea* ATPases all gave lines of complete identity, without spur development, even after prolonged incubation. ATPases from two strains of *Micrococcus roseus* and from *Micrococcus conglomeratus* and *Micrococcus varians* exhibited only partial cross reaction, as indicated by the presence of distinct spurs; in the latter case, spur formation was very weak. The ATPases from the membranes of *Micrococcus caseolyticus, Micrococcus rhodochrous, Sporosarcina ureae,* and *Bacillus subtilis* showed no reaction. When the enzymes from membranes of strongly cross-reacting species (*M. tetragenus, S. flava,* and *S. lutea*) were purified and electrophoresed along with purified ATPase from *M. lysodeikticus,* all exhibited apparently identical anodal electrophoretic mobilities, when developed against the *M. lysodeikticus* anti-ATPase serum, which indicated that the sizes and surface charges of these enzymes were very similar, if not the same. In contrast, the weakly and partially cross-reacting ATPase from *M. varians* exhibited a slightly different electrophoretic mobility. The results of the enzyme inhibition assays, by use of the different membrane ATPases with the antisera to *M. lysodeikticus* ATPase, strongly supported and expanded the results obtained in the gel diffusion and immunoelectrophoresis studies, because the antisera inhibited their enzymatic activities to different extents. The ATPases of strongly cross-reacting cocci closely related to *M. lysodeikticus* were almost as effectively inhibited as *M. lysodeikticus* ATPase, whereas weakly cross-reacting ATPases were inhibited to a lesser extent, and the ATPases from unrelated organisms that showed no cross reactivity were not inhibited to any extent (see Table 1). Whiteside et al (109) also compared the serologic relationships with a series of other independent parameters, including the quantitative and qualitative composition of the bacterial membrane phospholipids and fatty acids, the guanine and cytosine contents (mol % G + C) in the deoxyribonucleic acids of the species studied, and the types of cross bridges found in the cell wall peptidoglycans from these cocci. It was noted that all of these additional biochemical parameters correlated quite closely with the serologic cross reactivity of the various ATPases. The relationships of *M. lysodeikticus* anti-ATPase inhibition and cross reactivities in gels with the G + C contents of the other organisms are illustrated in Table 1. It was noted by these investigators (109) that the closer the taxonomic relationship between the different cocci, the greater was the similarity in the biochemistry and serology of their membrane ATPases.

4.7   Localization of Membrane ATPase by Immunoelectron Microscopy

As discussed earlier when observed under the electron microscope, negatively stained bacterial protoplast membranes reveal the presence of a large number of particles on the membrane surface (24,25,84). The alignment of such particulate structures along the edge of membrane fragments was very similar to the stalked particles observed on the inner membranes of mitochondria by Stiles and Crane (110) and Racker (6). The association of ATPase enzymatic activity with these bacterial membrane particles was first suggested by Muñoz et al. (84), who, along with Nachbar and Salton (8,111), noticed that the loss of these particles from the membrane of *M. lysodeikticus* was

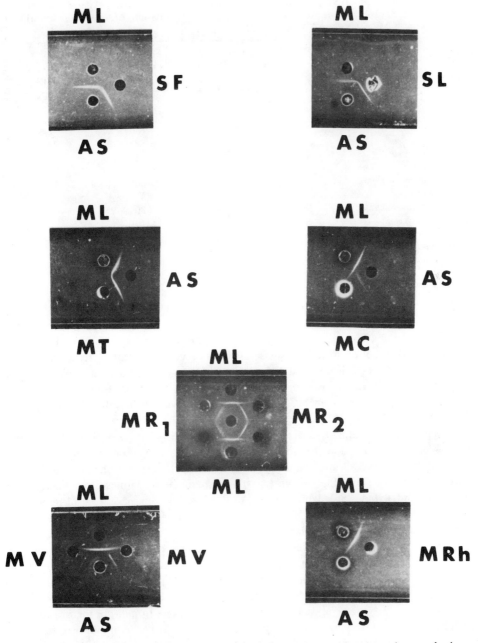

Fig. 7. Reactions of ATPase partially purified from the membranes of various pigmented micrococci with the antiserum to *M. lysodeikticus* soluble ATPase. The anti-ATPase serum (AS) was reacted in double-diffusion agar plates with partially purified ATPase from membrane of *S. flava* (SF), *S. lutea* (SL), *M. tetragenus* (MT), *M. conglomeratus* (MC), two strains of *M. roseus* (MR₁ and MR₂), *M. varians* (MV), *M. rhodochrous* (MRh), and also with membrane ATPase from *M. lysodeikticus* (ML) as the homologous control. From Whiteside et al. (109). By permission of *Journal of Bacteriology*.

Table 1. Relationship Between Cross Reactivity in Immunodiffusion Agar Gels, Enzyme Inhibition of Solubilized Membrane ATPase with Anti-ATPase to *M. lysodeikticus*, and G + C Contents of Micrococci and Other Bacteria[a]

Microorganism	Serological cross reactivity in diffusion gels	Inhibition by excess antibody (%)	G + C contents (mol%)
Micrococcus lysodeikticus		100	73.3
Micrococcus tetragenus	+	89	73.4
Sarcina flava	+	87	72.4
Micrococcus roseus	±	84	73.5
Micrococcus roseus R27	±	79	—
Sarcina lutea	+	75	73.0
Micrococcus conglomeratus	±	36	69.4
Micrococcus varians	−	35	72.4
Corynebacterium coelicolor	−	30	65.3
Micrococcus rhodochrous	−	0	70.4
Micrococcus caseolyticus	−	0	44.4
Sporosarcina ureae	−	0	42.9
Bacillus subtilis	−	0	43.0

[a] Data from Whiteside et al. (109).

associated with the simultaneous release of the enzyme and that in the solubilized form, the free enzyme looked strikingly similar to the membrane-bound particles. The particulate nature of purified ATPase isolated from several different bacteria, as observed by Muñoz et al. (84), Ishida and Mizushima (85), and Schnebli et al. (86), and their similarity to purified ATPases of mitochondrial origin, as seen by Kagawa and Racker (112) and Stiles and Crane (110), provided further evidence for the identity of these particles.

To specifically localize the enzyme, determine its distribution, and firmly establish that the particles that stud the isolated membranes of a variety of biologic cells are, in fact, ATPases, several investigators have attempted to use cytochemical techniques specific for the identification of the enzyme in combination with electron microscopy. Marchesi and Palade (56), who utilized ATP-dependent deposition of lead phosphate, clearly established the asymmetric distribution of erythrocyte membrane ATPase on the inner cytoplasmic face of the plasma membrane. Voelz (113) and Voelz and Ortigoza (31), by use of similar techniques, also showed that a bacterial ATPase was located on the cytoplasmic membrane, but in these studies, no firm conclusions about an asymmetric distribution could be reached. The major problem in both the erythrocyte and bacterial studies was that though the methods used were highly specific for a particular enzymatic activity, the resolution obtained was inadequate to indicate the precise location and number of enzymatic sites on these membranes. As has been indicated earlier in this review, the only way to achieve such exact resolution is by the combination of immunologic techniques with electron microscopy by the now classic methodology of immunoelectron microscopy. The use of such an approach to identify ATPase on the membranes of *M. lysodeikticus* has recently been reported by Oppenheim and Salton (23) and will be the topic of discussion in the following paragraphs.

Two essential prerequisites for the immunoelectron microscopic localization of any surface antigen by use of immunoferritin labeling are the specificity of the antibody employed in this technique and the ability of the antibodies to react with its eliciting antigen, when either or both are in a possibly altered state (e.g., soluble and membrane bound). The specificity of the antibody is obviously needed to ensure that the labeling reaction is restricted to only one of a possible number of antigens on a multiantigenic structure, such as a bacterial membrane, and therefore a homogeneous preparation of that antigen is a necessity. The specific antibody subsequently produced must then be able to react with the antigen, even when the antigen may be in a physical state different from that which exists at the time it was used to induce the antibody (i.e., antibody to the soluble form of a membrane component must be able to react with the membrane-bound form of that component in a detectable way) and after the antibody itself may be somewhat altered during the conjugation procedure. By utilizing the antiserum prepared against a homogeneous purified preparation of ATPase from *M. lysodeikticus* that was determined to successfully inhibit membrane-bound enzymatic activity (see Fig. 6) and that upon conjugation with ferritin was still shown to possess the ability to react with ATPase, Oppenheim and Salton (23) have firmly established the efficacy of the immunoferritin techniques in the localization of this enzyme on the protoplast plasma membranes.

The purified γ-globulin fraction of the anti-ATPase serum was conjugated to ferritin by the method of Singer and Schick (114). After separating out unreacted ferritin and γ-globulin and subsequently testing the conjugate by immunodiffusion assay to establish its reactivity with ATPase, the conjugate was used in the labeling of purified plasma membranes of *M. lysodeikticus*. The specificity of the labeling of membrane-bound ATPase with the antibody-ferritin conjugate can be visualized in electron micrographs of both negatively stained (Fig. 8) and thin-section preparations (Fig. 9) of labeled (Figs. 8B and 9B) and unlabeled (Figs. 8A and 9A) membranes. The distribution of ferritin conjugate in the negatively stained membranes is readily apparent and is strikingly similar to the patterns observed for the ATPase-associated particles on the untreated membranes. In many instances, ferritin molecules can be seen in direct contact with the ATPase-like structures (Fig. 8,B). From the thin-section studies, it is obvious that the specific ferritin anti-ATPase conjugate reacts with only one side of the membrane (Fig. 9, B), thus further indicating the asymmetric disposition of the enzyme. That the ATPase occurs on only the inner face of the bacterial plasma membrane, as initially indicated by the earlier studies of Fukui and coworkers (14) and Salton et al. (95), was confirmed by Oppenheim and Salton (23) by the inability to label the outer surface of intact protoplasts when such protoplasts were reacted with the conjugate. It is interesting to note in Fig. 9 that the membranes of *M. lysodeikticus*, in common with those from higher organisms, have a marked propensity toward vesicularization upon extended manipulations necessary for electron microscope examination, yielding both inside-out and right-side-out vesicles. The labeling of the outer surface of the vesicles seen in Fig. 9, in contrast to the absence of significant labeling of the intact protoplasts, strongly suggests that many of the vesicles must have turned inside-out with respect to surface orientation. Thus, the vesicles in Fig. 9,B that do not exhibit ferritin labeling either lack the ATPase or the enzyme may be inaccessible for reaction in vesicles that are right-side out and have closed prior to labeling.

Besides relying on visual verification of the labeling procedure, Oppenheim and Salton (23) also monitored labeling by immunologic and biochemical methods. Labeling was deemed successful when, after repeated washes of the conjugate-treated membranes, the ferritin was not released into the wash supernatants but was still detectable immunologically on the membranes in agar diffusion slides when labeled membranes were

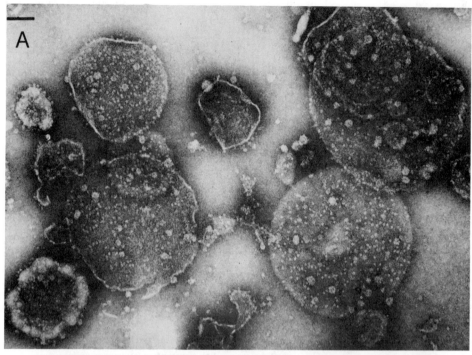

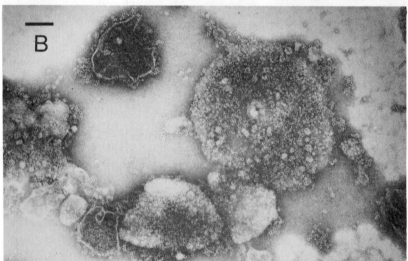

Fig. 8. Localization of ATPase particles on membranes from *M. lysodeikticus* by use of ferritin conjugated to antibody specific for the ATPase is illustrated in preparations that have been negatively stained with ammonium molybdate and examined in the electron microscope. Untreated, washed membranes in electron micrograph A show the presence of the uniform 10-nm diameter particles believed to be the ATPase. Labeling of the ATPase with the specific ferritin-antibody conjugate is illustrated in B. The bars in each preparation represent 0.1 μm. From Oppenheim and Salton (23). By permission of *Biochimica et Biophysica Acta.*

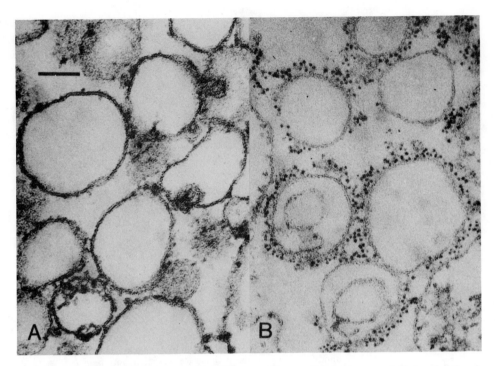

Fig. 9. The asymmetric distribution of ATPase particles on the membranes of *M. lysodeikticus* is shown in thin sections of (A) unlabeled preparation and membranes (B) labeled with the ferritin-anti-ATPase conjugate. Many of the vesicles show the inside-out orientation, with labeling on one face of the membrane, a result similar to that obtained with the cytochemical localization of erythrocyte membrane ATPase by Marchesi and Palade (56). The bar represents 0.1 μm. From Oppenheim and Salton (23). By permission of *Biochimica et Biophysica Acta*.

reacted against antiferritin sera (Fig. 10,A and B). It is interesting to note that while ferritin and, to a limited extent, the anti-ATPase γ-globulin to which it was conjugated, could be readily detected on the labeled membranes, the ATPase on treated or untreated membranes (Fig. 10, B1 and B2) could not be detected in agar gel diffusion tests by its homologous antisera. This may be due to the relatively limited accessibility of the membrane-bound form of the enzyme and/or because the steric arrangement of anti-ATPase antibody, though able to react with the enzyme, may not be in a proper configuration to cross-bridge membrane fragments into visible precipitates. On the other hand, the antibody conjugated to the ferritin and, to a greater extent, the ferritin itself are "sticking out" from the labeled membrane and are therefore much more amenable to react with their corresponding antiserum to yield a precipitin band in diffusion gels.

The other criterion Oppenheim and Salton (23) used for successful labeling was the subsequent inhibition of ATPase activity on the membrane after the reaction with the anti-ATPase ferritin conjugate. Detectable ATPase activity was observed to drop at least 90% after labeling. Residual enzymatic activity was attributed to sequestered ATPase that was not exposed during the labeling procedure (e.g., in closed right-side-out vesicles) but that became exposed during the washings and/or steric hindrance of some enzyme sites to reaction with antibody in the membrane-bound state. A summation of the enzymatic and immunologic results of a single labeling experiment can be seen in Table 2.

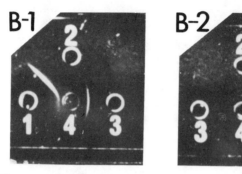

Fig. 10. Immunodiffusion reaction that shows the distribution of ferritin in three successive washes (wells A, B, and C) and on the residual *M. lysodeikticus* membrane preparation (well D) after ferritin labeling as detected by antiferritin antisera (well E). The reactions of ferritin-labeled membranes (well 4 of B-1) and unlabeled membranes (well 4 of B-2) against sheep antirabbit γ-globulin, anti-ATPase sera, and antiferritin serum were placed in wells 1–3, respectively, in B-1 and B-2. From Oppenheim and salton (23). By permission of *Biochimica et Biophysica Acta.*

Table 2. Distribution of ATPase Enzymatic Activity and Immunologically Detectable Ferritin, ATPase, and Anti-ATPase γ-Globulin in Membrane Fractions and Washes of Ferritin-labeled *M. lysodeikticus* Membranes

Sample	ATPase units	Ferritin[b]	ATPase[c]	Anti-ATPase[c] γ-globulin
Membrane suspension	+++[a]	−	−	−
Wash I	0	+++	−	++
Wash II	0	±	−	±
Wash III	0	−	−	−
Labeled membrane	±	++	−	+
Control membrane	+++[a]	−	−	−

[a] Indicates (+→+++) relative amount present or (−) absence of trypsin-activated ATPase; due to the complexity of kinetics of trypsin activation, exact units are not given.

[b] Indicates (+→+++) relative amount present or (−) absence of ferritin as detected in immunodiffusion assay (see Fig. 10, A and B).

[c] Indicates (+→+++) relative amount present or (−) absence of ATPase and anti-ATPase γ-globulin as detected in immunodiffusion assay (see Fig. 10, B). Data from Oppenheim and Salton (23).

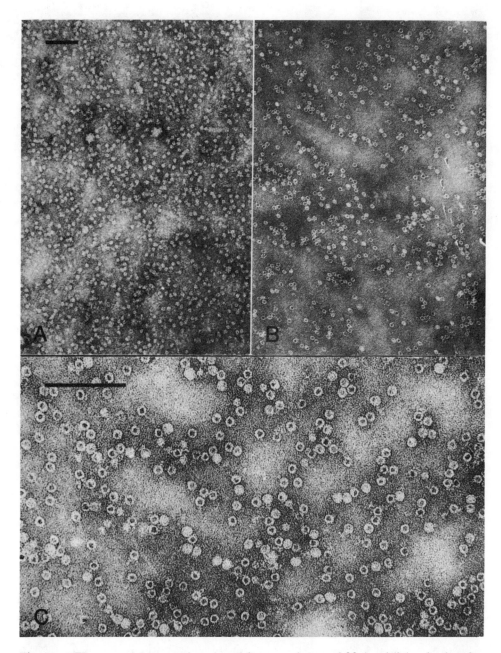

Fig. 11. ATPase-containing particles released from membranes of *M. lysodeikticus* by the selective shock-wash procedure are illustrated in the negatively stained preparation (A). Membranes that had been labeled with the ferritin-anti-ATPase conjugate, washed to remove unreacted label, and then subjected to the shock-wash procedure released the ferritin-antibody-ATPase complex, as shown in B and C. The bars represent 0.1 μm. From Oppenheim and Salton (23). By permission of *Biochimica et Biophysica Acta*.

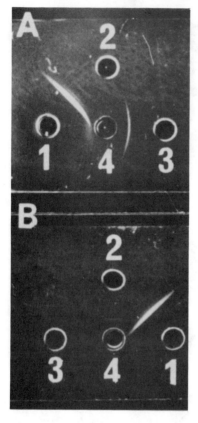
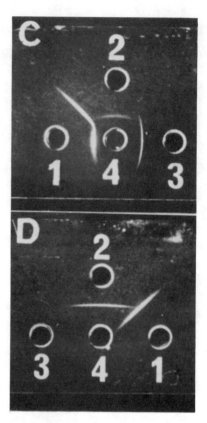

Fig. 12. The distribution of ATPase, ferritin, and anti-ATPase γ-globulin in various *M. lysodeikticus* membrane and shock-washed fractions as determined by immunodiffusion assays. Labeled membranes, unlabeled membrane, shock-washed material from the labeled membrane and the shock-wash from unlabeled membrane were respectively placed in the wells (4) on each of four separate slides (A–D). Wells 1–3 on each slide contained sheep antirabbit γ-globulin, rabbit anti-ATPase γ-globulin, and sheep antihorse ferritin antisera, respectively. From Oppenheim and Salton (23). By permission of *Biochimica et Biophysica Acta.*

In control experiments, Oppenheim and Salton (23) firmly established the specificity of the labeling procedure. Membranes treated with free ferritin, ferritin conjugated to nonimmune γ-globulin, or ferritin conjugated with specific antisera but blocked with excess antigen (i.e., ATPase) prior to labeling failed to show any significant attachment of ferritin conjugate to the membranes by electron microscopy or loss of enzymatic activity after such treatments. Likewise, membranes treated with an excess of antiserum or membranes that had been subjected to the shock-wash procedure for the release of ATPase prior to exposure to the conjugate showed little or no labeling, and, as expected, practically no enzymatic activity could be detected on these membranes. Attempts to label purified mesosomal membrane preparations prepared from intact protoplasts were also unsuccessful. The inability to achieve labeling was not surprising, in that ATPase was undetectable by immunologic or enzymatic (direct or trypsin activation) assays with purified or sonically disrupted mesosomal material.

As discussed earlier, Muñoz et al. (63,67) determined that the Ca^{2+}-activated ATPase can be released from the plasma membranes of *M. lysodeikticus* by the "shock-wash" or selective release procedure. Figure 11,A shows a micrograph of a typical negatively stained ATPase preparation obtained after such treatment, which consists primarily of the 10-nm diameter spherical particles associated with enzymatic activity and some small membrane fragments. When Oppenheim and Salton (23) subjected ferritin-labeled membranes (i.e., ferritin-anti-ATPase) to the same selective release procedure, ferritin-antibody-ATPase complexes were released from the membrane. These complexes can be visualized in negatively stained preparations, as seen in the low- and high-magnification electron micrographs in Fig. 11,B and C, where one or two ferritin molecules per ATPase molecule are observed. As expected from the reaction of specific antibody with soluble ATPase, such complexes were enzymatically inactive. When tested in immunodiffusion gels, however, each component of the complex (i.e., ferritin, anti-ATPase γ-globulin, and ATPase) was immunologically identifiable when reacted against its homologous antiserum, as can be seen in Fig. 12. Immunoelectrophoresis of the complex and control preparations of normal shock-released ATPase and free ferritin, developed against anti-ATPase γ-globulin and antiferritin, as seen in Fig. 13, further support the electron microscope and immunodiffusion studies and clearly establish that the ATPase was an integral part of the complex. Table 3 summarizes the distribution of ATPase enzymatic activity and immunologically detectable ATPase, anti-ATPase γ-globulin, and ferritin in the various fractions in a shock-wash treatment experiment of unlabeled and ferritin conjugate-labeled membranes. It should be pointed out that shocked ferritin-labeled membranes looked strikingly similar to untreated shocked membranes in the electron microscope, showing the attachment of very few residual ferritin or ATPase molecules, and that such membranes were unreactive with antifer-

Table 3. Distribution of ATPase Enzymatic Activity and Immunologically Detectable Ferritin, ATPase, and Rabbit Anti-ATPase γ-globulin in the Various Fractions in a Shock-wash Treatment Experiment of Unlabeled and Ferritin Conjugate-labeled *M. lysodeikticus* Membranes

Sample	ATPase units	Ferritin[b]	ATPase[b]	Rabbit Anti-ATPase[b]
Membranes	$+++^a$	−	−	−
Shock-wash of membranes	47.0	−	+++	−
Shock-washed membranes	$-^a$	−	−	−
Ferritin-labeled membranes	$\pm^a$	+++	−	+
Shock-wash of ferritin-labeled membranes	6.6	+++	+++	+++
Shock-washed ferritin-labeled membranes	$\pm^a$	−	−	−

[a] Indicates $(+\rightarrow+++)$ relative amount present or $(-)$ absence of trypsin-activated ATPase; due to the complexity of kinetics of trypsin activation, exact units are not given.

[b] Indicates $(+\rightarrow+++)$ relative amount present or $(-)$ absence of ferritin, ATPase, and anti-ATPase γ-globulin as detected in immunodiffusion assay (see Fig. 12). Data are from Oppenheim and Salton (23).

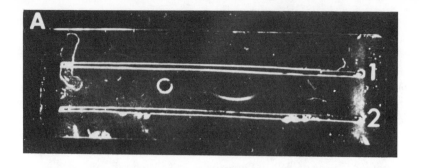

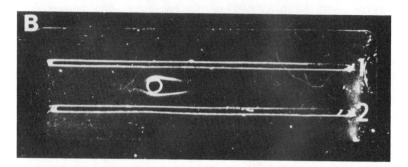

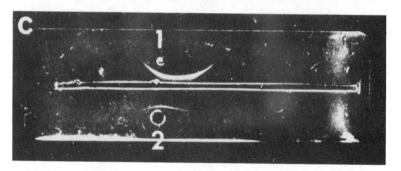

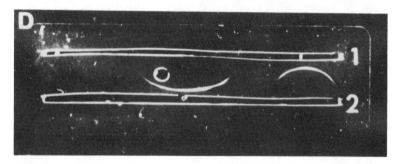

Fig. 13. Immunoelectrophoresis slide tests on fractions obtained by shock-wash treatment of unlabeled and ferritin-labeled membranes reacted against various antisera. A: Shock-wash from untreated membranes (well) electrophoresed and reacted against sheep antihorse ferritin antisera (trough 1) and rabbit anti-ATPase (trough 2). B: Shock-wash of ferritin-labeled membranes (well) electrophoresed and then reacted against antiferritin (trough 1) and anti-ATPase (trough 2). C: Purified, unconjugated ferritin (well 1) and shock-wash I of ferritin-labeled membranes (well 2) electrophoresed and then reacted against antiferritin (trough). D: Mixture of purified ferritin and shock-wash I from untreated membranes (center well) electrophoresed and then reacted against anti-ATPase (trough 1) and antiferritin (trough 2). Slides A–C were electrophoresed for 90 min, while slide D was run for 120 min. From Oppenhim and Salton (23). By permission of *Biochimica et Biophysica Acta.*

ritin or anti-ATPase sera. All of these results firmly established that the membrane particles are the sites of ATPase activity in *M. lysodeikticus*.

5. CONCLUSIONS

The studies on the bacterial membrane ATPase described above, we believe, offer an
excellent model by which one can elucidate some of the uncertainties that veil the
structural and functional organization of the biologic plasma membrane. This ubiquitous and vital membrane component has been carefully isolated, purified to
homogeneity, and many of its physical and enzymatic characteristics have been firmly
established. While its function(s) within the bacterial cell has yet to be firmly
determined, there are strong indications that it participates in oxidative phosphorylation, energy transduction, and cation transport. The ability to purify the ATPase had
led to the production of monospecific antiserum against this component. The antiserum
so prepared has served as a valuable reagent in a variety of studies for establishing
ATPase as a major antigenic component of the bacterial membrane, in further defining
the properties of the enzyme itself, in taxonomic studies for the recognition of structural
and antigenic differences and similarities in bacterial membranes, and in the localization
of the enzyme in the membrane by both qualitative and quantitative means. The latter
has been achieved by the highly selective method of immunoelectron microscopy by use
of ferritin-labeled antibody. Through the use of this technique, Oppenheim and Salton
(23) have been able to specifically localize the enzyme with a high degree of resolution
and firmly establish its position on the inner surface of the bacterial plasma membrane.
The techniques they have developed for the selective release of labeled complexes from
membranes clearly establish structure-function relationships and are valuable new tools
for studying the precise molecular architecture of biomembranes.

   The use of such a multidisciplinary approach should be especially applicable to the
study of any number of other plasma membrane components. In the event that studies
are fruitful, structural and functional relationships among individual membrane
components may then be established. In this way, it is hoped that we shall be able to
elucidate and understand the multifunctional biologic membrane at the molecular level.

REFERENCES

1. M. R. J. Salton, *Annu. Rev. Microbiol.,* **21,** 417 (1967).
2. M. R. J. Salton, in J. B. G. Kwapinski, Ed., *Molecular Microbiology,* John Wiley and
    Sons, New York, 1974, p. 387.
3. M. R. J. Salton, in L. A. Manson, Ed., *Biomembranes,* Vol. 1, Plenum Press, New York,
    1971, p. 1.
4. M. S. Nachbar, J. D. Oppenheim, and F. Aull, *Amer. J. Med. Sci.,* **268,** 122 (1974).
5. S. J. Singer and G. L. Nicolson, *Science,* **175,** 720 (1972).
6. E. Racker, in E. Racker, Ed., *Membranes of Mitochondria and Chloroplasts,* Van Nos-
    trand Reinhold Co., New York, 1970, p. 127.
7. N. A. Machtiger and C. F. Fox, *Annu. Rev. Biochem.* **42,** 575 (1973).

8. M. S. Nachbar and M. R. J. Salton, in M. Blank, Ed., *Surface Chemistry of Biological Systems,* Plenum Press, New York, 1970, p. 175.

9. M. R. J. Salton, in A. H. Rose and D. W. Tempest, Eds., *Advances in Microbial Physiology,* Vol. 11, Academic Press, London, 1973, p. 213.

10. J. B. G. Kwapinski, in J. B. G. Kwapinski, Ed., *Research in Immunochemistry and Immunobiology,* Vol. 2, University Park Press, Baltimore, Md., 1972, p. 1.

11. M. Heidelberger, in J. B. G. Kwapinski, Ed., *Research in Immunochemistry and Immunobiology,* Vol. 3, University Park Press, Baltimore, Md., 1973, p. 1.

12. H. Nikaido, in L. Leive, Ed., *Bacterial Membranes and Walls,* Vol. 1, Marcel Dekker, New York, 1973, p. 131.

13. M. R. J. Salton, *Trans. N.Y. Acad. Sci. Ser. II,* **29,** 764 (1967).

14. Y. Fukui, M. S. Nachbar, and M. R. J. Salton, *J. Bacteriol.,* **105,** 86 (1971).

15. I. Kahane and S. Razin, *J. Bacteriol.,* **100,** 187 (1969).

16. M. R. J. Salton, M. D. Schmitt, and P. E. Trefts, *Biochem. Biophys. Res. Commun.,* **29,** 728 (1967).

17. S. Rottem and S. Razin, *J. Bacteriol.,* **94,** 359 (1967).

18. Y. Fukui and M. R. J. Salton, *Biochim. Biophys. Acta,* **288,** 65 (1972).

19. T. L. Whiteside and M. R. J. Salton, *Biochemistry,* **9,** 3034 (1970).

20. B. Cinader, in B. Cinader, Ed., *Antibodies to Biologically Active Molecules,* Pergamon Press, Oxford, 1967, p. 85.

21. E. D. Day, in J. B. G. Kwapinski, Ed., *Research in Immunochemistry and Immunobiology,* Vol. 3, University Park Press, Baltimore, Md., 1973, p. 41.

22. P. Owen and J. H. Freer, *Biochem. J.,* **129,** 907 (1972).

23. J. D. Oppenheim and M. R. J. Salton, *Biochim. Biophys. Acta,* **298,** 297 (1973).

24. D. Abram, *J. Bacteriol.,* **89,** 855 (1965).

25. V. I. Biryuzova, M. A. Lukoyanova, N. S. Gel'man, and A. I. Oparin, *Dokl. Akad. Nauk USSR,* **156,** 198 (1964).

26. D. Branton, *Proc. Nat. Acad. Sci. USA,* **55,** 1048 (1966).

27. D. Branton and D. Southworth, *Exp. Cell Res.,* **47,** 648 (1967).

28. M. R. J. Salton, *Crit. Rev. Microbiol.,* **1,** 161 (1971).

29. W. van Iterson and W. Leene, *J. Cell Biol.,* **20,** 361 (1964).

30. A. W. Sedar and M. R. Burde, *J. Cell Biol.,* **24,** 285 (1965).

31. H. Voelz and R. O. Ortigoza, *J. Bacteriol.,* **96,** 1359 (1968).

32. I. Nisonson, M. Tannenbaum, and H. C. Neu, *J. Bacteriol.,* **100,** 1083 (1969).

33. K. J. Cheng, J. M. Ingram, and J. W. Costerton, *J. Bacteriol.,* **107,** 325 (1971).

34. A. H. Coons, H. J. Creech, R. N. Jones, and E. Berliner, *J. Immunol.,* **45,** 159 (1942).

35. S. J. Singer, *Nature (London),* **183,** 1523 (1959).

36. U. Hämmerling, T. Aoki, E. de Harven, E. A. Boyse, and L. J. Old, *J. Exp. Med.,* **128,** 1461 (1968).

37. U. Hämmerling, T. Aoki, H. A. Wood, L. J. Old, E. A. Boyse, and E. de Harven, *Nature (London),* **223,** 1158 (1969).

38. T. Aoki, H. A. Wood, L. J. Old, E. A. Boyse, E. de Harven, M. P. Lardis, and C. W. Stackpole, *Virology,* **45,** 858 (1971).

39. M. E. Lamm, G. C. Koo, C. W. Stackpole, and U. Hämmerling, *Proc. Nat. Acad. Sci. USA,* **69,** 3732 (1972).

40. C. Neauport-Sautes, D. Silvestre, M.-G. Niccolai, F. M. Kourilsky, and J. P. Levy, *Immunology,* **22,** 833 (1972).

41. L. Wofsy, P. C. Baker, K. Thompson, J. Goodman, J. Kimura, and C. Henry, *J. Exp. Med.,* **140,** 523 (1974).

42. M. J. Karnovsky, E. R. Unanue, and M. Leventhal, *J. Exp. Med.,* **136,** 907 (1972).

43. G. L. Nicolson and S. J. Singer, *Proc. Nat. Acad. Sci. USA* **68,** 942 (1971).

44. C. W. Stackpole, T. Aoki, E. A. Boyse, L. J. Old, J. Lumpley-Frank, and E. de Harven, *Science,* **172,** 472 (1971).

45. P. K. Nakane, J. SriRam, and G. B. Pierce, *J. Histochem. Cytochem.,* **14,** 789 (1966).

46. P. K. Nakane and G. B. Pierce, *J. Histochem. Cytochem.,* **14,** 929 (1966).

47. S. Avrameas and J. Uriel, *C. R. Acad. Sci. Ser. D* **262,** 2543 (1966).

48. G. A. Andres, K. C. Hsu, and B. C. Seegal, in D. M. Weir, Ed., *Handbook of Experimental Immunology,* Blackwell, Edinburgh, 1967, p. 527.

49. C. Morgan, *Int. Rev. Cytol.,* **32,** 291 (1972).
50. M. Wagner, in J. B. G. Kwapinski, Ed., *Research in Immunochemistry and Immunobiology,* Vol. 3, University Park Press, Baltimore, Md., 1973, p. 185.
51. L. A. Sternberger, *Immunocytochemistry,* Prentice-Hall, Englewood Cliffs, N.J., 1974.
52. J. W. Shands, *Ann. N.Y. Acad. Sci.,* **133,** 292 (1966).
53. J. R. Coulter and T. M. Mukherjee, *Infect. Immunity,* **4,** 650 (1971).
54. P. K. Nakane, G. E. Nichoalds, and D. L. Oxender, *Science,* **161,** 182 (1968).
55. I. R. Pendleton, K. S. Kim, and A. W. Bernheimer, *J. Bacteriol.,* **110,** 722 (1972).
56. V. T. Marchesi and G. E. Palade, *J. Cell Biol.,* **35,** 385 (1967).
57. A. Askari and S. N. Rao, *Biochem. Biophys. Res. Commun.,* **36,** 631 (1969).
58. E. Racker, *Fed. Proc.,* **26,** 1335 (1967).
59. E. Racker and L. L. Horstman, *J. Biol. Chem.,* **242,** 2547 (1967).
60. A. Abrams, *J. Biol. Chem.,* **240,** 3675 (1965).
61. A. Abrams and C. Baron, *Biochemistry,* **6,** 225 (1967).
62. D. Evans, *J. Bacteriol.,* **100,** 914 (1969).
63. E. Muñoz, M. R. J. Salton, M. H. Ng, and M. T. Schor, *Eur. J. Biochem.,* **7,** 490 (1969).
64. A. Askari, Ed., Properties and Functions of $(Na^+ + K^+)$ Activated Adenosinetriphosphatase, *Ann. N.Y. Acad. Sci.,* **242** (1974).
65. C. Weibull, *J. Bacteriol.,* **66,** 688 (1953).
66. S. Ishikawa and A. L. Lehninger, *J. Biol. Chem.,* **237,** 2401 (1962).
67. E. Muñoz, M. S. Nachbar, M. T. Schor, and M. R. J. Salton, *Biochem. Biophys. Res. Commun.,* **32,** 529 (1968).
68. M. Ishida and S. Mizushima, *J. Biochem., Tokyo,* **66,** 33 (1969).
69. R. Mirsky and V. Barlow, *Biochim. Biophys. Acta,* **241,** 835 (1971).
70. A. Hachimori, N. Muramatsu, and Y. Nosoh, *Biochim. Biophys. Acta,* **206,** 426 (1970).
71. A. Kobayashi and Y. Anraku, *J. Biochem., Tokyo,* **71,** 387 (1972).
72. P. L. Davies and P. D. Bragg, *Biochim. Biophys. Acta,* **266,** 273 (1972).
73. M. P. Roisin and A. Kepes, *Biochim. Biophys. Acta,* **305,** 249 (1973).
74. J. Carreiva, A. J. Leal, M. Rojas, and E. Muñoz, *Biochim. Biophys. Acta,* **307,** 541 (1973).
75. N. Nelson, B. I. Kanner, and D. L. Gutnick, *Proc. Nat. Acad. Sci. USA,* **71,** 2720 (1974).
76. M. Futai, P. C., Steneveis, and L. A. Heppel, *Proc. Nat. Acad. Sci. USA,* **71,** 2725 (1974).
77. H. Monteil, J. Schoun, and M. Guidard, *Eur. J. Biochem.,* **41,** 525 (1974).
78. G. Weinbaum and R. Markman, *Biochim. Biophys. Acta,* **124,** 207 (1966).
79. D. J. Evans, Jr., *J. Bacteriol.,* **104,** 1203 (1970).
80. R. L. Hanson and E. P. Kennedy, *J. Bacteriol.,* **114,** 772 (1973).
81. M. R. J. Salton and M. T. Schor, *Biochem. Biophys. Res. Commun.,* **49,** 350 (1972).
82. M. R. J. Salton and M. T. Schor, *Biochim. Biophys. Acta,* **345,** 74 (1974).
83. M. T. Schor, M. C. Heincz, M. R. J. Salton, and F. Zaboretzky, *Microbios,* **10A,** 145 (1974).
84. E. Muñoz, J. H. Freer, D. Ellar, and M. R. J. Salton, *Biochim. Biophys. Acta,* **150,** 531 (1968).
85. M. Ishida and S. Mizushima, *J. Biochem., Tokyo,* **66,** 133 (1969).
86. H. P. Schnebli, A. E. Vatter, and A. Abrams, *J. Biol. Chem.,* **245,** 1122 (1970).
87. C. Baron and A. Abrams, *J. Biol. Chem.,* **246,** 1542 (1971).
88. E. Racker, *Biochem. Biophys. Res. Commun.,* **10,** 435 (1963).
89. F. M. Harold, *Bacteriol. Rev.,* **36,** 172 (1972).
90. F. M. Harold and D. Papineau, *J. Membrane Biol.,* **8,** 45 (1972).
91. J. D. Butlin, G. B. Cox, and F. Gibson, *Biochim. Biophys. Acta,* **292,** 366 (1973).
92. G. B. Cox and F. Gibson, *Biochim. Biophys. Acta,* **346,** 1 (1974).
93. Y. Fukui, M. S. Nachbar, and M. R. J. Salton, *Biochim. Biophys. Acta,* **241,** 30 (1971).
94. S. Yang and R. S. Criddle, *Biochemistry,* **9,** 3063 (1970).
95. M. R. J. Salton, M. T. Schor, and M. H. Ng, *Biochim. Biophys. Acta,* **290,** 408 (1972).
96. H. Monteil and B. Schreiber, *Ann. Microbiol. Inst. Pasteur,* **124A,** 193 (1973).
97. H. Monteil and G. Roussel, *Biochem. Biophys. Res. Commun.,* **63,** 313 (1975).
98. G. Mancini, J. P. Vaerman, A. O. Carbonara, and J. F. Heremans, *Protides Biol. Fluids Proc. Colloq.,* **11,** 370 (1964).
99. B. Cinader, *Ann. N.Y. Acad. Sci.,* **103,** 495 (1963).
100. R. Arnon, *Eur. J. Biochem.,* **5,** 583 (1968).

101. G. Schatz, H. S. Penefsky, and E. Racker, *J. Biol. Chem.*, **242,** 2552 (1967).
102. J. D. Butlin, G. B. Cox, and F. Gibson, *Biochem. J.*, **124,** 75 (1971).
103. G. B. Cox, N. A. Newton, J. D. Butlin, and F. Gibson, *Biochem. J.*, **125,** 489 (1971).
104. B. J. Kanner and D. C. Gutnick, *FEBS Lett.*, **22,** 197 (1972).
105. S. Rottem and S. Razin, *J. Bacteriol.*, **92,** 714 (1966).
106. M. Argaman and S. Razin, *J. Gen. Microbiol.*, **55,** 45 (1969).
107. R. Y. Stanier, D. Wachter, C. Gasser, and A. C. Wilson, *J. Bacteriol.*, **103,** 387 (1970).
108. A. C. Wilson and N. O. Kaplan, in C. A. Leone, Ed., *Taxonomic Biochemistry and Serology,* Ronald Press, New York, 1964, p. 321.
109. T. L. Whiteside, A. J. De Siervo, and M. R. J. Salton, *J. Bacteriol.*, **105,** 957 (1971).
110. J. W. Stiles and F. L. Crane, *Biochim. Biophys. Acta,* **126,** 179 (1966).
111. M. S. Nachbar and M. R. J. Salton, *Biochim. Biophys. Acta,* **223,** 309 (1970).
112. Y. Kagawa and E. Racker, *J. Biol. Chem.,* **241,** 2461 (1966).
113. H. Voelz, *J. Bacteriol.,* **88,** 1196 (1964).
114. S. J. Singer and A. F. Schick, *J. Biophys. Biochem. Cytol.,* **9,** 519 (1961).

CHAPTER 5

STRUCTURAL AND FUNCTIONAL RELATIONSHIPS OF FATTY ACID SYNTHETASES FROM VARIOUS TISSUES AND SPECIES

STUART SMITH

The enzymes concerned with the *de novo* biosynthesis of fatty acids exist in nature in two distinct forms that have been classified as type I and type II (1). The type-I fatty acid synthetases are high-molecular-weight multienzyme complexes in which the component enzymes are tightly bound together as a single unit. The component enzymes of the type-II synthetase systems function jointly as a unit but are not bound together physically and exist as discrete proteins. Fatty acid synthetases of yeasts, fungi, and animals are all of the type-I variety, and until comparatively recently, it was generally assumed that bacteria and plants employed only the type-II synthetase systems. However, with the discovery of type-I multienzyme complexes in *Mycobacterium phlei* (1), *Corynebacterium diphtheriae* (2), *Streptomyces coelicolor* (3), and *Streptomyces erythreus* (4), it now seems that, in fact, type-I systems may be relatively common among the more advanced prokaryotes.

Supported by Grants AM 16073, AM 17489, and GRS SO1 RR05467 from the National Institutes of Health and Grant BMS 7412723 from the National Science Foundation. Dr. Smith is an Established Investigator of the American Heart Association.

Table 1. Properties of Type-I FASs

	Corynebacterium diphtheriae	Mycobacterium phlei	Streptomyces erythreus	Penicillium patulum	Saccharomyces cerevisiae	Euglena gracilis (etiolated)	Plaice liver	Pigeon liver
Native enzyme								
Molecular *mass* (daltons × 10^{-6})	2.75	1.70	>1.0	2.3	2.3	1.7	0.48	0.45
$s_{20,w}$ (sec × 10^{13})		37.2			40.6		12.6	14.0
$D_{20,w}$ (cm²sec⁻¹ × 10^7)								2.50
Stokes radius (Å)		108						82.6
Partial specific volume								0.744
Frictional ratio								1.62
SH groups per mole molecule								62–63
4'-phosphopantheine residues per mole molecule					3			1
pH optimum				7.6			6.5–6.6	6.7
Activation energy (kcal/mol)								
Primer specificity	$C_2 > C_4$							$C_2 > C_4$
Major products		C_{16}, C_{18}		C_{16}, C_{18}	C_{16}, C_{18}		C_{16}, C_{18}	C_{16}
		C_{24} CoAs	C_{16} CoA	CoAs	CoAs	C_{16} CoA	Acid	Acid
Electron donor	NADH	NADH			NADH	NADH		
	NADPH	NADPH			NADPH	NADPH	NADPH	NADPH
FMN requirement	Yes	Yes	Yes		Yes			No
K_m NADPH (μM)	6	2			4–7	2	19	9
K_m NADH (μM)	2.5	3			5	2		
K_m acetyl CoA (μM)	25			6.7		3.5	70	2.1
K_m malonyl CoA (μM)	10			7.2		45.5	60	3
Subunit species								
Molecular *mass* (daltons × 10^6)								0.22
$s_{20,w}$ (sec × 10^{13})								9.6
$D_{20,w}$ (cm² sec⁻¹ × 10^7)								4.13
Stokes radius (Å)								51
Frictional ratio								1.27
Reference	2	1, 7	4	8	9	5, 11	12	13–17

a Demassieux and Lachance (54).

The discovery of type-I synthetases in members of the genus *Streptomyces* is of interest from the taxonomic standpoint, because these organisms possess a bacterium-type cell wall but grow in the form of mycelia (4). The presence of type-I synthetases in *Streptomyces* lends support to the theory that these organisms are rather more closely related to yeasts and fungi than to bacteria.

A type-I synthetase has been found in *Euglena gracilis* (5) but not in *Chlamydomonas reinhardi* (6), a related phytoflagellate. The type-I synthetase is found in *E. gracilis* only when the organism is grown in the dark, and the organism apparently switches to a type-II system when exposed to light. This phenomenon is probably related to the bimodal life-style of *E. gracilis*, because this organism is phototrophic in the light but heterotrophic in the dark. In contrast, *C. reinhardi* is exclusively phototrophic and employs a type-II system, both in the light and dark.

The type-I synthetases found in primitive organisms are quite distinct from those found in higher animals. A comparison of the properties of the type-I enzymes from various sources is presented in Table 1. The type-I synthetases found in primitive organisms are usually of molecular weight 1,700,000–2,300,000, require both flavin

Chicken liver	Mouse liver	Mouse mammary	Rat liver	Rat mammary	Rabbit liver	Rabbit mammary	Guinea pig mammary	Pig liver	Cow mammary	Human liver
0.50	0.586	0.594	0.474	0.478	0.450	0.91 (0.487[a])	0.40	0.50	0.53	0.41
11.8			12.3	12.9	14.0	16.5 (13.8[a])	12.3		13.5	12.0
2.44				2.56	2.58	1.6	2.8		2.12	
86.1				84.6	83	(80.4[a])			96.4	
0.74				0.74	0.73					
1.63				1.63	1.63	(1.54[a])			1.74	
70				56	83	58–59 (62–56[a])			78	
			1	1						1
6.7			6.8	6.5–6.8	6.7	6.6	6.6			
29				28						
			$C_4 > C_2$	$C_4 > C_2$	$C_4 > C_2$				$C_4 > C_2$	$C_4 > C_2$
C_{16}	C_{16}	C_{16}	C_{16}	C_{16}	C_{16}	C_{16}	C_{16}	C_{16}, C_{18}	C_{16}	C_{16}
Acid	Acid	Acid	Acid	Acid	Acid		Acid	Acid	Acid	Acid
NADPH	NADPH	NADPH	NADPH	NADPH	NADPH	NADPH	NADPH	NADPH	NADPH	NADPH
			No 59							
					44	34			14	
				22	17	9 (24[a])	24			
				13	37	29 (14[a])	27			
0.253				0.225		(0.22[a])			0.22	
8.46				8.9					9.7	
3.24				4.28					3.95	
67.0				49					53.5	
1.59				1.18				26	1.34	
18, 19	20	20	20, 21	22, 23	24	25		27, 28	29	30

mononucleotide and NADPH for optimal activity, and release the reaction products as coenzyme A (CoA) esters. In contrast, the type-I systems of higher animals are usually of molecular weight ~500,000, require only NADPH, and release the reaction products as free acids.

It is evident from examination of Table 1 that a considerable amount of interest has been shown in the fatty acid synthetase systems of lactating mammary glands. The reason for this interest is that the lactating mammary glands of many species have the ability to synthesize medium-chain-length fatty acids that are not produced by other tissues. This point is illustrated in Table 2 for the rat. Whereas slices of the lactating mammary gland incorporated large amounts of labeled glucose into fatty acids of chain length C_8–C_{14}, slices of liver incorporated label almost exclusively into fatty acids with 16 and 18 carbon atoms. Studies with homogenate fractions from lactating mammary glands have demonstrated that these tissues synthesize the medium-chain-length fatty acids by the *de novo* malonyl CoA pathway (31). Consequently, investigators have isolated and characterized fatty acid synthetase multienzyme complexes from the lactating mammary glands of several species in the hope of learning something of the mechanism of chain termination in fatty acid biosynthesis. It is not uncommon to find in different tissues multiple forms of enzymes that catalyze the same basic reaction. These enzymes may have different kinetic properties, they may be structurally related, as in the case of some of the hybrid isoenzymes, or they may be structurally unrelated, as in the case of the liver and muscle fructose-1,6-diphosphatases (32). The existence of multiple molecular forms of enzymes could contribute to the versatility of organisms, providing molecular configurations most appropriate for the most specialized metabolic requirements. However, as can be seen from Table 1, no obvious differences in the physicochemical properties of the mammary and liver enzymes were revealed. Furthermore, all of the animal fatty acid synthetases (FASs), when incubated under optimum cofactor conditions, synthesized the same products, mainly palmitic acid, irrespective of the tissue of origin of the enzyme. It seemed conceivable that the FAS multienzyme complex isolated from mammary glands might not be the enzyme responsible for the synthesis of the medium-chain-length fatty acids *in vivo*. Therefore, the possibility that a separate FAS was required for the synthesis of these fatty acids was examined. By making use of rabbit antibodies raised against the purified FAS multienzyme complex from rat mammary gland, we were able to remove the FAS multienzyme complex from the cytosol of lactating mammary gland and examine the effect on the chain length of fatty

Table 2. Chain Length of Fatty Acids Synthesized from [U-^{14}C]glucose by Slices of Lactating Rat Liver and Lactating Mammary Gland[a]

Tissue	Glucose incorporated (nmol/g/hr)	% Distribution of radioactivity in fatty acids								
		$C_{8:0}$	$C_{10:0}$	$C_{12:0}$	$C_{14:0}$	$C_{16:0}$	$C_{16:1}$	$C_{18:0}$	$C_{18:1}$	Others
Liver	789	0.7	1.7	0.6	8.7	44.6	6.4	19.0	15.8	2.5
Mammary	4070	5.7	28.8	29.2	18.6	15.5	0.5	0.8	0.5	0.4

[a] Livers were obtained from young adult male rats that had been fasted 2 days and refed a fatfree high-glucose diet for 3 days. Mammary glands were obtained from lactating rats, 8 days postpartum.
Slices (100 mg) were incubated in 1 ml of Krebs-Henseleit bicarbonate buffer (pH 7.3) that contained [U-^{14}C]glucose (25 mM) for 3 hr at 37°C, with 95% O_2, 5% CO_2 as gas phase. Data are from Smith and Abraham (56).

IMMUNOELECTROPHORESIS

A

B

A

IMMUNODIFFUSION

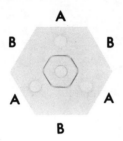

A

B   B

A   A

B

Fig. 1. Immunoelectrophoresis and Ouchterlony immunodiffusion analysis of fatty acid synthetase. Sample wells labeled A contained cytosol from lactating rat mammary gland equivalent to 5.9 units of FAS activity. Sample wells labeled B contained 5.9 units of purified rat mammary gland FAS. Rabbit antibodies against rat mammary gland FAS were placed in the trough for immunoelectrophoresis and in the center well for immunodiffusion. From Smith and Abraham (23). By permission of *The Journal of Biological Chemistry*.

acids synthesized (23). First, we had to establish that the antibodies were highly specific in their action. The reaction of the antibodies with cytosol and purified FAS from lactating rat mammary gland was studied. Immunoelectrophoresis of the cytosol and purified FAS gave a single precipitin line, the same distance from the antigen well (Fig. 1). Similarly, Ouchterlony immunodiffusion analysis with cytosol and purified FAS placed in adjacent wells each gave a single precipitin line that fused to form a regular hexagon (Fig. 1). These experiments established that the antibodies prepared against the purified FAS multienzyme complex reacted with a single component in the cytosol from lactating mammary gland. We next confirmed that the antibodies specifically inhibited the FAS activity in the mammary gland cytosol and had no effect on the activity of other enzymes involved in the conversion of acetate to fatty acid, that is, the acetate activating enzyme and acetyl CoA carboxylase (Fig. 2). Having established to our satisfaction that the antibodies were highly specific in their action, we were able to remove the FAS multienzyme complex from the mammary gland cytosol and look at the effect of the deletion on the chain length of the fatty acids synthesized.

Portions of cytosol from lactating rat mammary gland were pretreated with various quantities of antibody and then incubated with the cofactors required for the conversion

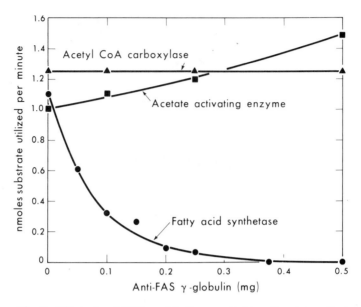

Fig. 2. Effect of anti-FAS γ-globulin on activities of acetate-activating enzyme, acetyl CoA carboxylase, and fatty acid synthetase in cytosol from lactating rat mammary gland. From Smith and Abraham (23). By permission of *The Journal of Biological Chemistry.*

of labeled acetate to fatty acid (Fig. 3). Synthesis of all C_{10}, C_{12}, C_{14}, and C_{16} chain-length fatty acids was completely inhibited in the cytosol that had been pretreated with antibodies. Thus, it was firmly established that the FAS complex was, indeed, an obligatory enzyme for the synthesis of all chain length fatty acids by the cytosol of lactating rat mammary gland. The possibility that a different synthetase was responsible for synthesis of medium-chain-length fatty acids was ruled out.

Our lack of success in unveiling some unique property of the mammary gland FASs that might provide a clue as to their unique mechanism of action *in vivo* raised the question of whether the FASs from mammary gland and other tissues were identical enzymes. This problem prompted a comparative survey of the FASs of various animals, with particular reference to the synthetases from different tissues of the same species.

1. MOLECULAR WEIGHT

The FASs isolated in our laboratory from rat liver and mammary gland appeared to have very similar molecular weights (474,000 and 478,000, respectively; Fig. 4, as did the FASs from mouse liver and mammary gland (mol wt 586,000 and 594,000, respectively; Fig. 5). Rabbit liver (24) and mammary gland (25) FASs, which are reported to have molecular weights of 450,000 and 910,000, respectively, would appear to be exceptions to this trend, and this point will be discussed later in the chapter.

2. QUATERNARY STRUCTURE

The FASs from rat liver and mammary gland are both cold labile enzymes that undergo reversible cold-induced dissociation into their half-molecular-weight subunits

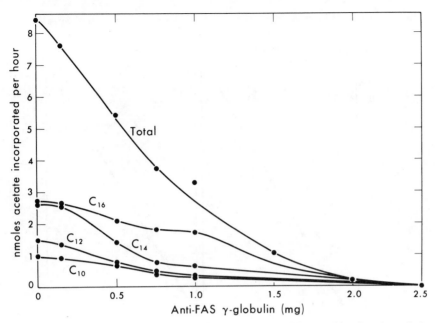

Fig. 3. Effect of anti-FAS γ-globulin on the synthesis of fatty acids of various chain length by cytosol from lactating rat mammary gland. Cytosol (0.2 mg of protein) was preincubated with antibodies against lactating mammary gland FAS, and fatty acid synthesis was then estimated by measuring the incorporation of labeled acetate. From Smith and Abraham (23). By permission of *The Journal of Biological Chemistry*.

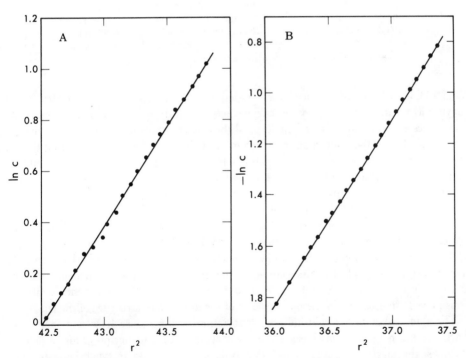

Fig. 4. Sedimentation equilibrium of purified FASs from (A) lactating rat mammary gland (1 mg/ml) and (B) liver (0.5 mg/ml). The calculated molecular weights were 478,000 and 474,000 for the mammary and liver enzymes, respectively. From Smith and Abraham (22) and Smith (20). By permission of *The Journal of Biological Chemistry* and *Archives of Biochemistry and Biophysics,* respectively.

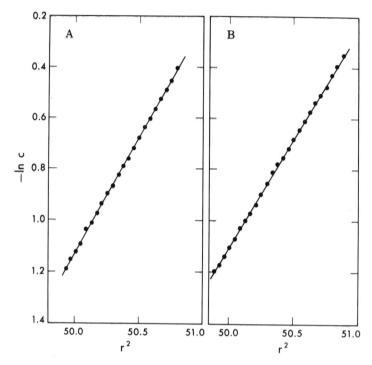

Fig. 5. Sedimentation equilibrium of purified FASs from (A) mouse liver and (B) lactating mammary gland. Initial protein concentrations were 0.5 mg/ml. Calculated molecular weights were 586,000 and 594,000 for the liver and mammary enzymes, respectively. From Smith (20). By permission of *Archives of Biochemistry and Biophysics*.

(20,22,33,34). This cold-induced dissociation of the rat liver and mammary gland enzymes is accompanied by a loss in enzyme activity. Both the dissociation and inactivation processes follow first-order kinetics, and the rate constants for the reactions are very similar for the rat liver and mammary enzymes (Table 3). The observation that the rat mammary gland enzyme was stabilized by high-ionic-strength media (22) suggested that perhaps hydrophobic interactions were important in the maintenance of the native form of the enzyme. Because hydrophobic interactions are thought to be stabilized in ^2H_2O media, the effect of ^2H_2O on the stability of the enzyme was studied. The rate constants for cold-induced dissociation and inactivation of the enzyme were identical and were decreased to similar extents by the presence of ^2H_2O (33). Because the subunits of the FAS are themselves inactive (34), it was concluded that the cold-induced inactivation was a direct result of the dissociation of the enzyme into the component subunits (33,34). Furthermore, the stabilizing effect of ^2H_2O confirmed that hydrophobic interactions played a major role in the maintenance of the quaternary structure of the enzyme.

The rate constants for cold-induced dissociation of the mouse liver and mammary FASs are much lower than those of the rat enzymes (20). Furthermore, for the mouse enzymes, the inactivation and dissociation processes do not occur synchronously (35). These observations suggest that whereas the liver and mammary gland enzymes from the same species show very similar stability characteristics, distinct differences exist in the quaternary structure of the rat and mouse enzymes. In particular, hydrophobic interactions appear to play a relatively less important role in the case of the mouse enzymes.

The FASs from pigeon liver (36), chicken liver (18,19), guinea pig mammary gland (26), rabbit liver (24), and rabbit mammary gland (25) [but not cow mammary gland (29)] all undergo inactivation in the cold, but the process is not paralleled by dissociation into subunits in all cases. Rather less attention has been paid to study of the stability of the FASs from more primitive organisms. *S. erythreus* (4) and *C. diphtheriae* (2) FASs, in common with the FASs from animal sources, are stabilized by high-ionic-strength media, which suggests that hydrophobic interactions may be of importance in the maintenance of the integrity of these FASs too.

3. AMINO ACID COMPOSITION

In Table 4, the amino acid compositions of several FASs are presented. The composition of the yeast enzyme shows a number of distinct differences compared to the enzymes from animal sources. The rat liver and mammary enzymes appear to be almost identical, as do the rabbit liver and mammary enzymes, with the possible exception of their lysine content. The chicken and pigeon liver enzymes also are very similar. Only a few minor differences are evident among the bird, rabbit, and rat enzymes. The amino acid compositions of some proteins, such as cytochrome *c* and various pyridine nucleotide-requiring dehydrogenases (38), insulin, and immunoglobulins (39), have been successfully correlated with their structure, function, and even evolution. It is possible that with very-large-molecular-weight proteins, such as the FASs, amino acid composition may not provide a very sensitive index of structural similarity because of the possibility of internal compensation.

4. IMMUNOLOGIC CROSS REACTIVITY

It has been demonstrated that the immunologic cross reactivity of homologous proteins is closely related to their sequence resemblance (40,41). This relationship can be demonstrated when cross reactivity is compared by any of the usual procedures, such as immunodiffusion, quantitative precipitin reactions, or microcomplement fixation. It has been demonstrated that changes in the conformation of proteins can significantly influence their antigenic reactivity (42), so that immunologic correspondence is probably more a reflection of the similarity of the surfaces of cross-reacting proteins.

TABLE 3. Dissociation of FASs into Subunits

FAS	First order rate constants (days)$^{-1}$	
	0°C	20°C
Rat mammary gland	0.4	0.03
Rat liver	0.3	0.02
Mouse mammary gland	0.02	0
Mouse liver	0.03	0

[a] From Smith (20).

Table 4. Amino Acid Compositions of Various FASs

Amino acid	mol%[a]							
	Ye[b]	ChL	PiL	RbL	RbM	RaL	RaM	CoM
Lys	6.87	5.15	5.35	2.52	4.47	4.27	4.30	5.19
His	1.77	2.71	2.84	2.55	2.46	2.77	3.10	2.71
Arg	3.92	4.16	4.90	5.93	5.96	4.83	5.23	4.31
Asp	11.03	8.91	9.03	8.14	8.24	8.32	8.33	8.67
Thr	6.32	5.24	4.86	5.32	4.65	5.25	5.49	5.11
Ser	6.76	8.03	6.88	7.13	6.23	7.26	8.02	7.08
Glu	12.42	11.85	12.22	11.81	12.45	10.66	11.39	14.28
Pro	5.01	5.33	4.86	7.86	7.47	5.99	6.39	7.53
Gly	7.78	8.47	8.51	9.01	8.59	8.88	8.47	8.09
Ala	8.01	8.21	8.87	11.30	10.87	9.52	9.12	7.82
Val	7.27	8.01	8.51	7.18	7.89	8.11	7.50	6.79
Ile	6.32	4.89	4.89	2.39	3.33	4.04	3.79	4.18
Leu	8.89	12.52	12.06	13.49	12.35	14.43	12.86	12.42
Tyr	3.17	3.11	2.70	2.06	1.75	2.24	2.45	2.53
Phe	4.45	3.41	3.52	3.31	3.28	3.42	3.54	3.30
Reference	9	19	16	24	25	20	22	37

[a] Because incomplete data are available for some enzymes and because of the uncertainty as to the exact molecular weight of some enzymes, the data have not been expressed as residues per mole of enzyme.

[b] Ye, yeast; ChL, chicken liver; PiL, pigeon liver; RbL, rabbit liver; RbM, rabbit mammary; RaL, rat liver; RaM, rat mammary; CoM, cow mammary.

The cross reactivity of FASs from a wide variety of species representative of fishes, amphibians, reptiles, birds, and mammals has been compared in several laboratories, and the data are collected in this chapter. The first studies to be reviewed here were performed with rabbit antisera prepared against FAS from the lactating rat mammary gland. The cross reactivity of partially purified FASs from pigeon, mouse, and rat liver and from rat, rabbit, and mouse mammary gland were compared by immunodiffusion and immunoelectrophoresis. The pigeon liver and rabbit mammary enzymes gave no precipitin line in the immunodiffusion and immunoelectrophoretic experiments (Figs. 6 and 7). Both the liver and mammary gland FASs of the rat and mouse gave precipitin lines in the immunodiffusion test, with some evidence of spur formation when rat and mouse enzymes were placed in adjacent wells (Fig. 6). These enzymes also gave precipitin lines in the immunoelectrophoresis experiments, the lines forming approximately the same distance from the antigen well (Figs. 7 and 8). The effect of the anti- (rat mammary) FAS γ-globulin on the activity of several mammary gland FASs was also compared. The activities of the mammary gland FASs of the rat and mouse were inhibited to similar extents by the antibodies, but the rabbit enzyme was relatively unaffected (Fig. 9).

In the immunodiffusion study with highly purified antigens (Fig. 10,A), reactions of complete identity were obtained among the rat liver, mammary gland, and adipose FASs. Similarly, reactions of complete identity were observed with the mouse liver and mammary gland enzymes (Fig. 10,B). All possible combinations of the five homogeneous antigens in adjacent wells are shown in Fig. 10,C and 10,D. Whenever FASs from different species were present in adjacent wells, reactions of partial identity were

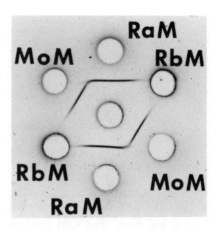

A   **B**

Fig. 6. Immunodiffusion analyses of partially purified FASs. A: Outer wells contained 1 unit of FAS activity. PiL, pigeon liver; MoL, mouse liver; RaL, rat liver. B: Outer wells contained 4 units of FAS activity. RaM, rat mammary gland; RbM, rabbit mammary gland; MoM, mouse mammary gland. Center wells contained 0.25 mg of rabbit anti- (rat mammary gland) FAS γ-globulin. From Smith (20). By permission of *Archives of Biochemistry and Biophysics*.

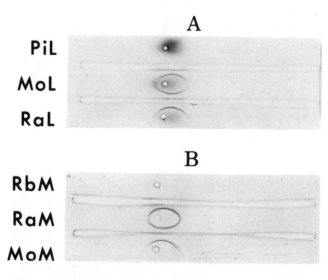

Fig. 7. Immunoelectrophoresis of various FASs. A: Partially purified preparations of FASs (1 unit) were placed in the sample wells. B: Cytosols that contained 4.1 units of FAS activity were used as the source of FAS. In both experiments, the troughs contained rabbit anti- (rat mammary) FAS γ-globulin.

135

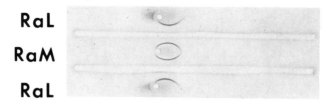

Fig. 8. Immunoelectrophoresis of FASs prepared from rat liver and lactating mammary gland. Sample wells contained cytosol equivalent to 2.3 units of FAS activity; troughs contained rabbit anti- (rat mammary) FAS γ-globulin.

obtained. The spur formation always pointed toward the well that contained the mouse enzymes, indicating they were the least related to the parent antigen, that is, the rat mammary gland FAS. Immunodiffusion analysis of the FAS purified from mouse mammary gland tumors showed this enzyme to be indistinguishable from the FAS from normal lactating mouse mammary gland (Fig. 11).

The immunodiffusion test, although convenient for distinguishing between proteins of different cross reactivity, is somewhat insensitive to small antigenic differences. Prager and Wilson (41) have shown that the quantitative immunoprecipitin test is rather more sensitive. Their studies with various lysozymes demonstrated that the threshold for the formation of spurs, as opposed to lines of complete identity, corresponded to about 80% cross reactivity in the quantitative immunoprecipitin test. Therefore, to determine whether any small differences in cross reactivity could be detected between enzymes that gave lines of complete identity, the quantitative immunoprecipitin test was employed.

The immunoprecipitin titration curves of highly purified rat liver, mammary, and adipose FASs followed very similar profiles. The results of a typical experiment are presented in Fig. 12. The titration curves for the mouse liver and mammary enzymes also showed similar profiles but reached an equivalence point at a rather higher antigen : antibody ratio than did the rat enzymes. The composition of the precipitates also differed between mouse and rat enzymes. At equivalence point, the precipitates

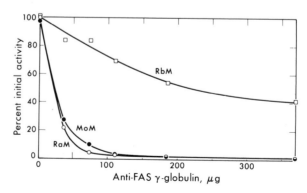

Fig. 9. Effect of rabbit anti- (rat mammary) FAS γ-globulin on the activity of various FASs. Cytosol fractions (1.6 units of FAS activity) were used as the source of FASs, and inhibition was determined by measuring the incorporation of [1,3^{14}C]malonyl CoA into fatty acid.

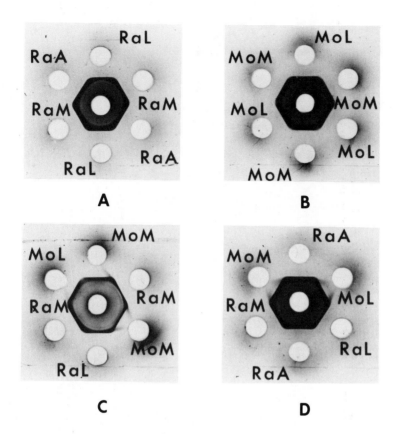

Fig. 10. Immunodiffusion analyses of homogeneous FASs. Outer wells contained 45 μg of purified FAS. Center wells contained 0.32 mg of rabbit anti- (rat mammary gland) FAS γ-globulin. From Smith (20). By permission of *Archives of Biochemistry and Biophysics*.

contained 1.6–1.8 μg of antibody/μg of antigen with the rat FASs but only 1.0–1.1 μg of antibody/μg of antigen with the mouse enzymes. Similar experiments were performed with different antisera raised against the rat mammary and liver FASs, and no difference between FASs from the same species could be detected.

Volpe et al. (43) have compared the cross reactivity in the immunodiffusion test of FASs from rat liver and brain. By use of an anti- (rat liver) FAS serum, they observed a reaction of complete identity (i.e., no spurring) between the two enzymes. A quantitative immunoprecipitin reaction also gave identical equivalence points for the rat brain and liver FASs (Fig. 13).

Yun and Hsu have conducted a survey of the cross reactivity of several FASs by use of the immunodiffusion test (19). Employing an anti- (chicken liver) FAS serum, they found that the pigeon and chicken liver enzymes gave reactions of partial identity, with the spur pointing to the well that contained the pigeon enzyme, as would be expected due to the fact that this enzyme was least related to the parent antigen (Fig. 14). No precipitin line was obtained with the FASs of rat and hamster liver. In some unpublished experiments, the same investigators extended their studies to include liver

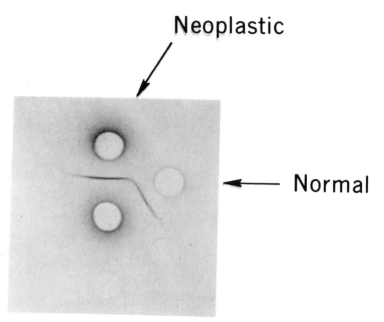

Fig. 11. Immunodiffusion of FASs from mammary tumor and normal lactating mammary gland of C_3H mice. The sample wells contained 17 units each of partially purified FAS. The antibodies used were raised against the FAS from lactating rat mammary gland. From Lin et al. (33). By permission of *Cancer Research*.

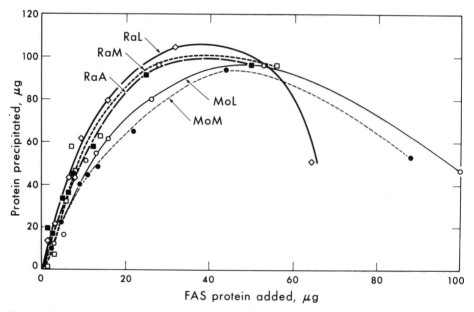

Fig. 12. Immunoprecipitin reactions of various FASs. The antibodies used were rabbit anti-(rat mammary) FAS γ-globulins. All antigen was precipitated up to equivalence point, as evidenced by the absence of FAS activity in the supernatant. From Smith (20). By permission of *Archives of Biochemistry and Biophysics*.

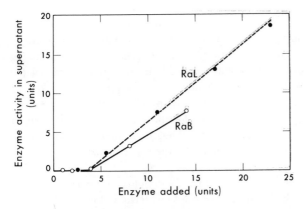

Fig. 13. Quantitative immunoprecipitin reactions of rat liver and brain FASs, with antibodies raised against rat liver FAS. From Volpe et al. (43) By permission of *The Journal of Biological Chemistry*.

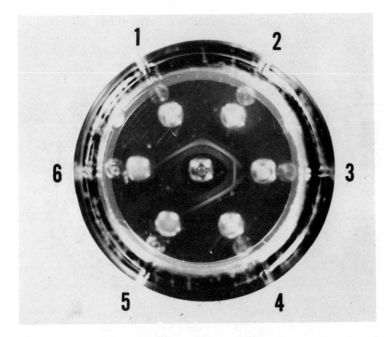

Fig. 14. Immunodiffusion analysis of various FASs. Outer wells 1, 2, and 4 contained crude chicken liver enzyme; wells 3, 5, and 6 contained crude enzyme from pigeon, rat, and hamster, respectively. Center well contained rabbit anti- (chicken liver) FAS serum. From Yun and Hsu (19). By permission of *The Journal of Biological Chemistry*.

FASs from bullhead fish, bullfrog, snake, and opossum. None of these enzymes formed precipitin lines with the anti- (chicken liver) FAS serum (Fig. 15). Interestingly, the turtle FAS did form a precipitin line of partial identity when adjacent to the pigeon or chicken enzymes. The spur always pointed toward the well that contained the turtle FAS when this enzyme was placed in wells adjacent to the pigeon or chicken enzymes. Because the FAS from snake did not form any precipitin line, the results can be interpreted to indicate that turtle FAS resembles the bird enzymes more closely than does that of its fellow reptile the snake. Clearly though, on the basis of the spur orientation, the pigeon enzyme is more clearly related to the chicken enzyme than is the turtle enzyme.

Recently, Carey in England has compared the immunologic cross reactivities of rabbit liver, mammary, and brain FASs, by use of the Ouchterlony double-diffusion test and the quantitative immunoprecipitin reactions. As I have pointed out earlier, according to the results of Carey and Dils (25), the rabbit mammary FAS is unique among all of the animal FASs studied thus far, in that it has a much higher molecular weight (Table 1). However, in view of the recent unpublished results of Demassieux and Lachance (Table 1), this point now seems to be in dispute. The rabbit liver FAS, on the other hand, appears to fall in line with the general trend of having a molecular weight in the 400,000–600,000 region. In view of the fact that all other studies have failed to demonstrate any differences between FASs from different tissues of the same species, Carey's experiments are particularly interesting. In the Ouchterlony immunodiffusion test, the partially purified rabbit liver, mammary, and brain enzymes gave single precipitin lines that fused completely without spur formation. The cow mammary gland FAS, also included in the study, gave no precipitin line. When quantitative

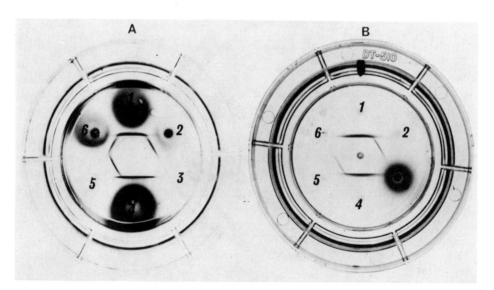

Fig. 15. Immunodiffusion analyses of various FASs. A: Outer wells 1 and 4 contained chicken liver cytosol (9.5 units); wells 2 and 6 contained pigeon liver enzyme (5 units); wells 3 and 5 contained partially purified turtle enzyme (5 units). B: Outer wells 1 and 4 contained purified chicken liver enzyme (6 units); wells 2 and 5 contained partially purified turtle enzyme (5 units); well 3 contained opossum liver cytosol (4 units); well 6 contained rat liver cytosol (8 units). The center wells contained rabbit anti- (chicken liver) FAS serum. From Yun and Hsu (58).

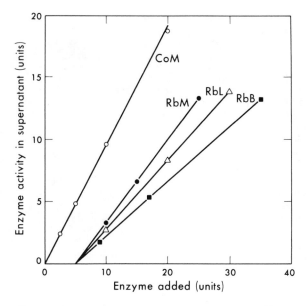

Fig. 16. Quantitative immunoprecipitin reactions of FASs from rabbit mammary gland, liver, and brain and lactating cow mammary gland. The rabbit FASs were partially purified enzymes, and the cow FAS was a homogeneous preparation provided by J. Knudsen. The antibodies were raised against the FAS from lactating rabbit mammary gland and used as a partially purified γ-globulin fraction. Based on unpublished data from E. Carey.

immunoprecipitin reactions were carried out with these enzymes, identical equivalence points were obtained with all three rabbit enzymes (Fig. 16). Because reactions of identity in both the immunodiffusion and quantitative immunoprecipitin tests almost certainly indicate an extremely high degree of sequence resemblance, it would seem that Demassieux and Lachance's estimate of the molecular weight of the rabbit mammary FAS is more consistent with the immunologic findings. The reason for the high molecular weight found by Carey and Dils is uncertain; conceivably, they may have been dealing with a "dimeric" form of the enzyme. However, Demassieux and Lachance (54) have homogenized rabbit mammary glands in both 0.25 M sucrose and isotonic potassium chloride and always obtained the "monomeric" form. They have also determined the molecular weight of the enzyme by molecular sieve chromatography and obtained a value of 460,000, in close agreement with the values they obtained by equilibrium sedimentation (487,000).

The results of the immunodiffusion tests performed in various laboratories have been summarized in Table 5. Clearly, the FASs from different tissues of the same species are always immunologically identical. This conclusion has been verified by the more sensitive quantitative immunoprecipitin test in all cases.

Comparison of the immunologic cross reactivity of homologous proteins from modern species can be regarded as an analysis of the degree of similarity retained from a common ancestral protein. Thus, biochemists have attempted to use data on cross reactivity of related proteins from different species to construct phylogenetic trees (44). While the

Table 5. Degree of Cross Reaction of Animal FASs in Ouchterlony Double-Diffusion Test

	Antisera[a]			
FAS	Rat mammary	Rat liver	Chicken liver	Rabbit mammary
Bullhead liver			0	
Frog liver			0	
Snake liver			0	
Turtle liver			PI	
Chicken liver			CI	
Pigeon liver	0		PI	
Opossum liver			0	
Rabbit liver				CI
Rabbit mammary	0			CI
Rabbit brain				CI
Hamster liver			0	
Mouse liver	PI			
Mouse mammary	PI			
Mouse mammary tumor	PI			
Rat liver	CI	CI	0	
Rat mammary	CI	CI		
Rat adipose	CI			
Rat brain		CI		
Cow mammary				0

[a]CI, complete identity; PI, partial identity; 0, no cross reaction.

results presented here are too incomplete to attempt such a task, it is interesting to note that the data do fit quite well into the classic phylogenetic schemes advanced by taxonomists (Fig. 17). Within the group of mammals studied, the rat and mouse FASs appear to be quite closely related, as one might expect, because they are both members of the Muridae group of rodents. Both the rat and mouse FASs are quite distinct from the rabbit enzyme, however. Although the rabbit was once thought to be closely related to the rodents, fossil records from the early Cenozoic era show that rodents and rabbits were already quite distinguishable from each other (45). Thus, the similarities of present-day rodents and rabbits may have resulted from convergent evolution. The rabbit, along with the hare and pika, is now classed in a separate order, the Lagomorpha. The FASs from birds, as might be expected, are quite distinct from those of fishes, amphibians, and both marsupial and placental mammals. However, the relationship of reptilian FASs is particularly interesting, in that the turtle enzyme appears to be more closely related to bird enzymes than is the FAS of its fellow reptile the snake.

Paleontologists tell us that the turtle has descended relatively unchanged from the early part of the Mesozoic era to the present day, whereas other reptiles have undergone much more extensive evolutionary changes during the last 200 million years (45). Thus, a relatively slower rate of evolutionary modification of the turtle FAS since the divergence of birds and reptiles could result in a greater resemblance to the common ancestrial enzyme than might be expected. It is interesting that the phylogenetic tree compiled by Fitch and Margoliash based on cytochrome c sequences also places the turtle closer to the birds than to its fellow reptiles (44). This anomaly illustrates that sequence or immunologic correspondence reflects not only the time elapsed since divergence but also the rate at which viable evolutionary modifications are made.

5. PRODUCT SPECIFICITY

The component of the FAS multienzyme that is responsible for termination of growth of the acyl chain is a thioesterase. The recent successful isolation of this thioesterase from the multienzyme by limited trypsinization of the complex (46,47) has enabled us to study in detail the properties of this key enzyme. The thioesterases isolated from the FAS complexes of both rat liver and mammary gland give reactions of complete identity when diffused with antibodies raised against the parent multienzyme. Furthermore, they have identical molecular weights, amino acid compositions, and kinetic properties, and both enzymes hydrolyze specifically long-chain thioesters (55). These observations provided the final conclusive evidence that regulation of the chain length of fatty acids synthesized *de novo* in the mammary gland could not be accounted for by properties inherent in the mammary gland FAS multienzyme.

The realization that the unique ability of mammary tissue to synthesize short- and medium-length fatty acids has not arisen through the evolution of a specialized FAS enzyme prompted a search for an auxiliary factor that might have the ability to modify

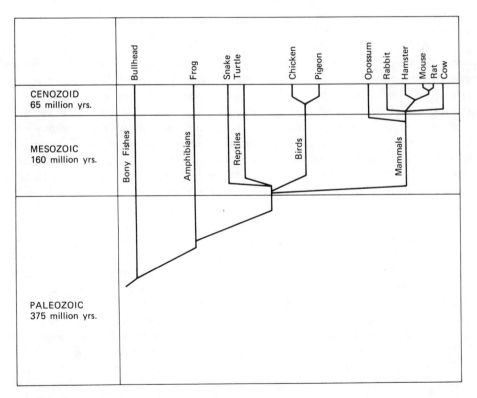

Fig. 17. Simplified phylogenetic tree that illustrates the probable evolutionary relationships of the vertebrates. Based largely on data from Romer (45).

Table 6. The Effect of a Component of Liver and Mammary Gland Cytosol on the Fatty Acids Synthesized by Purified Liver and Mammary Gland FASs[a]

No.	Cytosol extract	FAS	% Distribution of radioactivity					
			C_8	C_{10}	C_{12}	C_{14}	C_{16}	C_{18}
1	Liver	Liver	2.4	6.3	6.6	22.1	58.9	3.7
2	Liver	Mammary	8.6	5.0	8.1	20.9	52.6	4.8
3	Mammary	Liver	5.6	21.7	18.6	18.8	32.1	3.2
4	Mammary	Mammary	4.6	20.7	18.0	18.3	35.4	3.0

[a] The cytosol extracts were obtained by ammonium sulfate precipitation (35–60%). The concentration of cytosol extract and FAS added to incubations 1 and 2 was the same as that found in the intact liver cytosol. Similarly, the concentration of cytosol extract and FAS added to incubations 3 and 4 was the same as that found in the mammary cytosol. The cytosol extracts contained no detectable FAS activity. The products were determined by measuring the incorporation of [1-^{14}C]acetyl CoA, by use of the standard FAS assay system. From Smith (57).

product chain length. It has recently been shown by workers in Europe (48,49), and in our laboratory (50), that there is such a component present in the cytosol of lactating mammary glands. In Table 6 are shown the results of a "crossover" experiment in which cytosol extracts from liver and lactating mammary gland were incubated with purified FASs from liver and mammary gland. The fatty acid products synthesized by various combinations of these components were determined by use of [1-^{14}C]acetyl-CoA and unlabeled malonyl CoA as the substrates. Clearly, the product chain length was not influenced by the type of FAS added, as we would have expected on the basis that the liver and mammary FASs are indistinguishable. However, the source of the cytosol extract was instrumental in determining the product chain length. Thus, when a liver cytosol extract was used, the products were predominantly long chain, but when a cytosol extract from lactating mammary gland was used, significant amounts of medium-chain-length fatty acids were synthesized. The chain length modifier has recently been purified from lactating rat mammary gland by ammonium sulfate precipitation and chromatography on DEAE-cellulose and Sephadex G75 (51). The "chain-shortening" activity appears to copurify with a thioesterase activity. The thioesterase activity is associated with a protein of molecular weight 32,000, the same as that of the thioesterase component isolated from the FAS multienzyme. Although both thioesterases are also inactivated by phenylmethylsulfonyl fluoride, a serine esterase inhibitor, they are not identical enzymes and can be distinguished by both structural and functional criteria. The two thioesterases differ structurally, as evidenced by immunodiffusion studies. Whereas the thioesterase obtained from the FAS multienzyme (designated thioesterase I) reacts with antibodies raised against the FAS multienzyme complex, the other thioesterase (designated thioesterase II) does not. Functionally, the thioesterases differ in that thiosterase I is specific for long-chain thioesters, whereas thioesterase II has a broad specificity and will hydrolyze both medium- and long-chain thioesters. Thioesterase II is, nevertheless, a true thioesterase and has no activity toward carbonyl esters.

The properties of thioesterase II and its presence in lactating mammary gland strongly suggest that it is this enzyme that is responsible for modification of the product specificity of the FAS and, consequently, for the ability of the gland to synthesize the medium-chain-length fatty acids characteristic of milk fat.

Modification of the product specificity of an enzyme by a second component has been observed in the case of the bacterial ribonucleoside diphosphate reductase (52) and the mammalian lactose synthetase system (53). In the former case, the specificity for purine or pyrimidine is regulated by the binding of a nucleoside triphosphate to an effector site. The case of the lactose synthetase is rather different and in some ways may be analogous to the FAS system. The lactose synthetase system has two components, usually referred to as the A and B proteins. The A protein, a galactosyl transferase, is an enzyme in its own right, is widely distributed in animal tissues, and catalyzes the formation of N-acetyllactosamine from N-acetylglucosamine and uridine diphosphate (UDP) galactose. The B protein is identical with the milk protein α-lactalbumin but has no known enzymatic properties of its own. In the presence of the B protein, the specificity of the A protein is changed, so that the galactosyl moiety can be transferred from UDP-galactose to glucose rather than to N-acetylglucosamine. This change in specificity has been attributed to a marked increase in the K_m of the A protein for glucose. In the absence of the B protein, the K_m of the A protein for glucose is so high as to be prohibitive for the formation of lactose.

Like the lactose synthetase A protein, the FAS is found widely distributed in animal tissues. Its function is to synthesize long-chain fatty acids. In the lactating mammary gland, the function of the enzyme is rather different; that is, it synthesizes short- or medium-chain-length fatty acids. The altered specificity of the FAS in mammary gland is, like that of the lactose system, attributable to the presence of a second protein. The presence in lactating mammary gland of the second components of the lactose synthetase and FAS systems bestow on the gland the ability to perform certain tissue-specific processes, that is, the synthesis of lactose and medium-chain fatty acids. Perhaps through a fuller understanding of the mechanism of operation of these modifiers and their evolutionary origins, we will learn something of the general principles that relate to organ specificity and its development.

ACKNOWLEDGMENTS

I am grateful to Pauline Smith for her valuable help in the experimental part of this work and to Robert Hsu, Eric Carey, and Jean-Paul Lachance for sharing with me some of their unpublished observations.

REFERENCES

1. D. N. Brindley, S. Matsumura, and K. Bloch, *Nature (London)*, **224,** 666 (1969).
2. H. W. Knoche and K. E. Koths, *J. Biol. Chem.*, **248,** 3517 (1973).
3. M. Weinrich, unpublished results; cited in Ref. 2.
4. A. Rossi and J. W. Corran, *Biochem. Biophys. Res. Commun.*, **50,** 597 (1973).
5. I. Goldberg and K. Bloch, *J. Biol. Chem.*, **247,** 7349 (1972).
6. R. Sirevag and R. P. Levine, *J. Biol. Chem.*, **247,** 2586 (1972).
7. P. K. Flick and K. Bloch, *J. Biol. Chem.*, **249,** 1031 (1974).
8. K. Holtermuller, E. Ringelman, and F. Lynen, *Hoppe-Seyler's Z. Physiol. Chem.*, **351,** 1411 (1970).

9. F. Lynen, *Fed. Proc.*, **20**, 941 (1961).

10. H. B. White, O. Mitsuhashi, and K. Bloch, *J. Biol. Chem.*, **246**, 4751 (1971).

11. J. Delo, M. L. Ernst-Fonberg, and K. Bloch, *Arch. Biochem. Biophys.*, **143**, 384 (1971).

12. A. C. Wilson and I. P. Williamson, *Biochem. J.*, **117**, 268 (1970).

13. P. H. W. Butterworth, H. Baum, and J. W. Porter, *Arch. Biochem. Biophys.*, **118**, 716 (1967).

14. P. C. Yang, R. M. Bock, R. Y. Hsu, and J. W. Porter, *Biochim. Biophys. Acta*, **110**, 608 (1965).

15. E. J. Jacob, P. H. W. Butterworth, and J. W. Porter, *Arch. Biochem. Biophys.*, **124**, 392 (1968).

16. R. Y. Hsu, G. Wasson, and J. W. Porter, *J. Biol. Chem.*, **240**, 3736 (1965).

17. P. H. W. Butterworth, P. C. Yang, R. M. Bock, and J. W. Porter, *J. Biol. Chem.*, **242**, 3508 (1967).

18. R. Y. Hsu and S. L. Yun, *Biochemistry*, **9**, 239 (1970).

19. S. L. Yun and R. Y. Hsu, *J. Biol. Chem.*, **247**, 2689 (1972).

20. S. Smith, *Arch. Biochem. Biophys.*, **156**, 751 (1973).

21. D. N. Burton, A. G. Haavik, and J. W. Porter, *Arch. Biochem. Biophys.*, **126**, 141 (1968).

22. S. Smith and S. Abraham, *J. Biol. Chem.*, **245**, 3209 (1970).

23. S. Smith and S. Abraham, *J. Biol. Chem.*, **246**, 2537 (1971).

24. S. Demassieux and J. P. Lachance, *Biochim. Biophys. Acta*, **348**, 94 (1974).

25. E. M. Carey and R. Dils, *Biochim. Biophys. Acta*, **210**, 371 (1970).

26. C. R. Strong and R. Dils, *Int. J. Biochem.*, **3**, 369 (1972).

27. H. Dutler, M. J. Coon, A. Kull, H. Vogel, G. Waldvogel, and V. Prelog, *Eur. J. Biochem.*, **22**, 203 (1971).

28. H. Dutler, A. Kull, and R. Mislin, *Eur. J. Biochem.*, **22**, 213 (1971).

29. S. K. Maitra and S. Kumar, *J. Biol. Chem.*, **249**, 118 (1974).

30. D. A. K. Roncari, *Can. J. Biochem.*, **52**, 221 (1974).

31. S. Abraham, K. J. Matthes, and I. L. Chaikoff, *Biochim. Biophys. Acta*, **46**, 197 (1961).

32. M. Enser, S. Shapiro, and B. L. Horecker, *Arch. Biochem. Biophys.*, **129**, 377 (1969).

33. S. Smith, *Biochim. Biophys. Acta*, **251**, 477 (1971).

34. S. Smith and S. Abraham, *J. Biol. Chem.*, **246**, 6428 (1971).

35. C. Y. Lin, S. Smith, and S. Abraham, *Cancer Res.*, **35**, 3094 (1975).

36. S. Kumar, R. A. Muesing, and J. W. Porter, *J. Biol. Chem.*, **247**, 4749 (1972).

37. J. E. Kinsella, D. Bruns, and J. P. Infante, *Lipids*, **10**, 227 (1975).

38. T. P. Fondy and P. D. Holohan, *J. Theor. Biol.*, **31**, 229 (1971).

39. C. E. Harris and D. C. Teller, *J. Theor. Biol.*, **38**, 347 (1973).

40. E. M. Prager and A. C. Wilson, *J. Biol. Chem.*, **246**, 5978 (1971).

41. E. M. Prager and A. C. Wilson, *J. Biol. Chem.*, **246**, 7010 (1971).

42. M. Z. Atassi, *Biochim. Biophys. Acta*, **221**, 612 (1970).

43. J. J. Volpe, T. O. Lyles, D. A. Roncari, and P. R. Vagelos, *J. Biol. Chem.*, **248**, 2502 (1973).

44. W. M. Fitch and E. Margoliash, *Science*, **155**, 279 (1967).

45. A. S. Romer, *Vertebrate Palentology*, 2nd edit., University of Chicago Press, Chicago, 1945.

46. E. Agradi, L. Libertini, and S. Smith, *Biochem. Biophys. Res. Commun.*, **68**, 894 (1976).

47. S. Smith, E. Agradi, L. Libertini, and K. N. Dileepan, *Proc. Nat. Acad. Sci. USA*, **73**, 1184 (1976).

48. J. Knudsen and R. Dils, *Biochem. Biophys. Res. Commun.*, **63**, 780 (1975).

49. J. Knudsen, and S. Clark, and R. Dils, *Biochem. Biophys. Res. Commun.*, **65**, 921 (1975).

50. S. Smith and S. Abraham, *Advan. Lipid Res.*, **13**, 195 (1975).

51. L. Libertini, C. Y. Lin, and S. Smith, *Fed. Proc.*, **35**, 1609 (1976).

52. N. C. Brown and P. Reichard, *J. Mol. Biol.*, **46**, 39 (1969).

53. K. Brew, *Essays Biochem.*, **6**, 9 (1970).

54. S. Demassieux and J.-P. Lachance, Unpublished results. (1975).

55. S. Smith and C. Y. Lin, Unpublished results (1975).

56. S. Smith and S. Abraham, Unpublished results (1974).

57. S. Smith, Unpublished results (1974).

58. S. L. Yun and R. Y. Hsu, Unpublished results (1975).

CHAPTER 6

ENZYME ANALYSIS BY QUANTITATIVE IMMUNOELECTROPHORESIS

PETER OWEN AND CYRIL J. SMYTH

1.   INTRODUCTION

Quantitative immunoelectrophoresis refers to a group of methods that involve electrophoresis of antigens into an agarose gel that contains homogeneously distributed antibody, with or without prior separation of the anitgens in one of several supporting gel media. Rocket immunoelectrophoresis (RIE),* which has also been termed electroimmunoassay and electroimmunodiffusion, involves direct electrophoresis of antigen(s) into an agarose gel that contains monospecific antiserum or, occasionally, a polyvalent antiserum. Crossed immunoelectrophoresis (CIE), known also as two-dimensional immunoelectrophoresis, antigen-antibody crossed electrophoresis, two-dimensional (crossed) electroimmunodiffusion, and Laurell electrophoresis, comprises agarose gel electrophoresis in the first dimension, followed by electrophoresis at right angles into gels that contain antibody. Rocket- or arc-shaped immunoprecipitates are formed by the specific interaction of antigens with their corresponding antibodies. The surface areas delineated by individual immunoprecipitates can be approximated to peak heights for RIE and are proportional to the antigen-antibody ratio (1–5). Thus, these methods are suitable not only for immunochemical analysis and identification but also for quantitation of antigens and their corresponding antibodies. Moreover, the resolution achieved in the first dimension of two-dimensional variants of quantitative immunoelectrophoresis is retained, by contrast with the situation with diffusion-based immunoelectrophoretic procedures (6); for example, compare the results obtained by Fukui et al. (7) with those obtained later by Owen and Salton (8). This advantage has proved particularly important in the analysis of complex mixtures of antigens and also in the study of complex formation and microheterogeneity, where differences in electrophoretic mobility between complexes and free enzyme or between multiple forms of an enzyme are small.

  The purpose of this review is to discuss the contribution that quantitative immunoelectrophoretic procedures have made to the characterization, identification, and localization of enzymes since the pioneering methodologic studies of Laurell (9,10) 10 years ago. The methodology of quantitative immunoelectrophoresis and some of its many possible applications have been the subject of three monographs (11–13), two reviews (14,15), and a literature survey (5), all of which are recommended for complete

  The authors have been supported by Grant BMS 75-03934 from the National Science Foundation to M. R. J. Salton and Grant AI-10999 from the National Institutes of Health to A. E. Friedman-Kien.
  * CIE, crossed immunoelectrophoresis; CLIE, crossed-line immunoelectrophoresis; RIE, rocket immunoelectrophoresis; RLIE, rocket-line immunoelectrophoresis; ATP, adenosine triphosphate; DEAE, diethylaminoethyl; EDTA, ethylenediaminetetraacetic acid; MTT, tetrazolium-3-(4,5-dimethylthiazolyl-2)-2,5-diphenyltetrazolium bromide; NAD, nicotinamide adenine dinucleotide; NADH, **nicotinamide,** adenine dinucleotide reduced form; NDP, nucleotide diphosphate; NTP, nucleotide triphosphate; SDS, sodium dodecyl sulfate.

practical and theoretical instruction. Only methodologic aspects that are important enough to bear repetition or are of particular relevance to the scope of this chapter will be mentioned. Most of the results discussed herein were obtained by crossed immunoelectrophoresis or by two-dimensional variations of this procedure. However, although application of the newer methodologic modifications, such as sodium dodecyl sulfate (SDS) polyacrylamide gel crossed immunoelectrophoresis, to enzyme immuno-chemistry is limited, this aspect will be given some emphasis, because it should find increasing usage with the wider knowledge of its great potential. Similarly, the use of fused rocket immunoelectrophoresis will be stressed in relationship to the monitoring of fractionation procedures, because this method offers considerable advantages over classic methods. For example, quantitative immunoelectrophoresis techniques will, by and large, measure total enzyme content irrespective of activity expressed by the enzyme.

Localization of enzymes by CIE will be dealt with, firstly, in terms of solving problems of taxonomy and of tissue or membrane specificities, such as determination of membrane marker enzymes and detection of cytoplasmic contamination and, secondly, in terms of the molecular arrangement of enzymes on the membrane. Indeed, CIE has recently proved of exceptional value in determining the localization, with respect to the inner and outer surfaces of the plasma membrane, of five membrane-associated enzymes of *Micrococcus lysodeikticus* (8). However, problems do exist in interpreting immunoprecipitation patterns obtained when using detergent-solubilized membrane preparations. One of these problems, namely, the observation that some immunoprecipitates can possess more than one enzyme activity, will be dealt with in some detail, because the existence of multienzyme complexes in detergent extracts of membrane fractions may give us insight into some of the complex functional associations that exist within the cell membrane.

Later paragraphs will illustrate the importance of CIE in monitoring complex formation, such as that which occurs between proteolytic enzymes and acute phase globulins of serum, and, conversely, its potential in monitoring fragmentation and dissociation of antigens, such as occurs during the activation of prothrombin and during other reactions of the blood clotting process. The immunologic relationships between constituent subunits of enzyme molecules have also been successfully probed by use of CIE or related combination techniques, and these relationships will be discussed in a separate section on enzyme structure.

A detailed discussion of the use of CIE techniques in clinical chemistry and its application to the study of complement lysis are obviously outside the scope of this review, however. Consequently, mention will be made only of those problems of clinical importance that involve body fluid enzymes and components of nonhuman origin, such as bacteria.

2.   GENERAL METHODOLOGY

2.1   Antibodies

The use of a carefully standardized reference antiserum is obviously of considerable importance in any high-resolution analytic system. Thus, it is important to document the properties of the antiserum and antibodies used in these procedures. The degree of heterogeneity of an antiserum depends on the duration of the immunization period, the immunization schedule, the animal chosen for immunization, the frequency of bleedings, the administration of boosters, and the nature of the immunogen (15–19). The choice of animal species may be particularly important when investigating antigenic similarities and differences between native and modified or induced variants of mammalian enzymes (16,20). It should be remembered that animal sera may contain preimmune or natural antibodies to mammalian or bacterial antigens (21–24) and that animals may also produce antibodies that cross react with autologous proteins (18). Individual immune responses can vary considerably with identical administration of the same immunogen(s), especially with complex antigens, such as membranes or cytosol (20,25–28). To obviate problems associated with changing reference immunoprecipitation patterns caused by differences in antibody response with time and between animals, serum pools of large volume have been recommended for systematic investigations (15,17,27).

Production of monospecific antiserum depends to a large extent on the homogeneity of the initial immunogen. The problems associated with the purification of some antigens, such as sialoglycoproteins, may be circumvented by charge modification at a convenient point in the purification procedure. Thus, Brogren and colleagues have shown that desialylated serum cholinesterase is suitable for the production of antiserum monospecific for the native sialoprotein (29). Care, of course, is required in choosing a procedure that does not alter the immunologic properties of the antigen.

Electrophoresis of an enzyme into antibody-containing agarose gel requires that antibodies are immobile under conditions that permit electrophoretic migration of the antigen. Under normal experimental conditions, this situation is an idealized one, because antibodies are electrophoretically heterogeneous (2,5). Some enzymes, for example, subtilopeptidase (5,30) and trypsin (31–33), may interact with nonimmunoglobulin components in unfractionated antiserum. In such cases, purified immunoglobulins must be used. This has the additional advantage of greatly decreasing background staining of the agarose gel and increasing the contrast between immunoprecipitates and background (2). Purified immunoglobulin preparations, however, may exhibit proteolytic activity due to the presence of plasmin (34), which can cause degradation of proteins during electrophoresis into the antibody-containing gel. This phenomenon is responsible for the multiplicity of congruent immunoprecipitates obtained for spectrin (18,35–37,390). Such congruency has also been observed for one of the two NADH dehydrogenase isoenzymes from *Neisseria gonorrhoeae* envelopes (24) and for components of culture filtrates of *Clostridium tetani* (38). The degradative reactions can be prevented by the addition of protease inhibitors, such as the polypeptide aprotinin (Trasylol) to gels that contain antibody (34).

2.2 Antigens

One percent agarose gels possess exclusion limits for proteins of $\geq 10^8$ daltons and thus permit separation and analysis of antigens mainly on the basis of charge. Standardization of the antigen preparation is as important (5,14) as standardization of antiserum pools, because even antigens produced by similar procedures can yield inconsistent results (27).

The average mobility of any protein, including antibodies, in a quantitative immunoelectrophoretic system has an electrical and an electroendosmotic component; the latter is affected by many parameters (5,39). Factors that influence mobility include buffer strength, pH, buffer composition, agarose purity, buffer additives, such as detergents or divalent cations, viscosity additives, such as dextran 70,000, and the animal species of antibodies used (2,5,20,40). Basic antigens, such as subtilopeptidase (30) and IgG (41), may not be resolved in gels with medium to high electroendosmosis (39,42). Such effects can be alleviated by purification or charge modification of agarose (5), by selection of certain antibody populations to permit CIE at higher pH values (41), by use of antibodies from a different animal species (5), by treatment of antigen preparations with low concentrations of an ionic detergent, such as SDS (43) or 1,3-propiolactone (44), or, alternatively, by acetylation (45), carbamylation (46–48), formylation (49), citraconylation, or maleylation (50) of antigens.

To standardize the first-dimensional electrophoresis in CIE, bromphenol blue-albumin (22,51) and Evans blue-albumin (27) have been used as indices of migration. However, bromphenol blue should be avoided as a tracking dye, because it has been shown to interact with proteins, such as galactose dehydrogenase, to induce artifactual oligomeric multiple forms (52) and with others, such as dihydrofolate reductase, to cause microheterogeneity (53). For accurate comparison of antigens on the basis of relative mobility after CIE, carbamylated transferrin (46; see also Fig. 10), transferrin (54), albumin (17,55), and acetylated albumin (45) have been successfully employed as internal standard markers because of their high anodal mobility.

2.3 Quantitative Aspects and Sensitivity

As will be discussed more fully in a later section of this review, quantitative immunoelectrophoresis of enzymes measures not just functional concentrations but, rather, total antigenic content. Although this method is extremely sensitive, sample volume restrictions often require the use of antigen preparations concentrated to 5–10 mg of protein/ml. This is especially true when examining complex antigenic mixtures, such as tissue extracts or cytoplasmic or membrane fractions from microorganisms.

Accurate intragel quantitation by RIE and CIE requires homogeneous distribution of immunoglobulins throughout the antibody-containing gel (1,3). Comparability between gels is facilitated by the use of an internal standard, such as carbamylated transferrin (5). In RIE experiments in which peak height rather than peak area is used to estimate antigen content, deviation from linearity can occur when the antigen range is wide (56,57). This deviation may explain why Milisauskas and Rose (43,58) obtained linearity with semilogarithmic plots of antigen concentration against peak height for a prostatic acid phosphatase from urine and for several enzymes from human diploid (WI-38) cells. Cholinesterase (29), alanine aminopeptidase (59), lysozyme (60–62), pepsin (63), auxin oxidase (64), α-amylase (65), and subtilopeptidase (30) have also been quantitated by RIE.

Quantitation of individual antigens, such as barley amylases (66), by CIE can be performed by measuring the areas delineated by immunoprecipitates, by planimetry, or by other methods, including triangulation or weighing copies of the peaks made on tracing paper (4,5). Poor baseline definition (5,27,67), however, will not allow accurate area measurement. Analysis of polymodal peaks, such as those produced by multiple forms of an enzyme (68–71), could be tackled by curve analysis with a computer (72).

The sensitivity of quantitative immunoelectrophoresis differs slightly for individual proteins and depends on the limit of visual detection of stained immunoprecipitates. As little as 0.5–10 ng of protein antigen can be detected with Coomassie brilliant blue (2,3). Various enhancement techniques can be used to make weak immunoprecipitates visible. Thus, addition of dextrans or polyethylene glycol can enhance by sharpening the peaks of precipitation (73–75). Other additives intensify protein staining by further deposition of protein, such as antiimmunoglobulin (76,77) and guinea pig complement (78,79), or "mordants," such as tannic acid (5,80) and cadmium acetate (77). Radiolabeling methods, which involve soaking immunoplates in suitably labeled solutions, are generally more sensitive and have been used to detect thyroxine-binding proteins in serum (81) and antibodies to allergens (82). The combined use of radiolabeled antibodies and autoradiography can lower the detection limit of antigens to 10 pg of protein (83). In contrast, the limit of detection of human cholinesterase when enzyme staining techniques are used is about 100–150 pg (84). Peroxidase-conjugated antiimmunoglobulins have recently been used to reveal immunoprecipitates (402).

Basic enzymes, such as subtilopeptidase (30), the esterase and glucuronidase of human diploid cell lines (43), or auxin oxidases (64), can be conveniently quantitated by peak measurements on cathodal rockets. In other instances, carbamylation of the antigen may be necessary, as was the case for human carbonic anhydrases (47,48). Lowering the operating pH conditions together with carbamylation of the antibodies (85) may be particularly useful with enzymes that are irreversibly inactivated at the alkaline pH commonly used (pH 8.2–8.6).

Reversed rocket immunoelectrophoresis (87,391), that is, RIE of antibody into antigen-containing agarose, has permitted quantitation of the immune response to an enzyme, for example, to diphtheria toxin, a nicotinamide adenine dinucleotidase (NADase) that catalyzes adenosine diphosphate ribosylation of elongation factor 2 (88,89).

2.4 Detergents and Membrane Antigens

The preparation of various mammalian and microbial membranes has recently been comprehensively reviewed (90), as has detergent extraction of membrane proteins (91). The principal aim in CIE analysis of membrane-bound enzymes is to achieve "solubilization" with concomitant retention of both immunologic specificity and enzymatic activity. The choice of detergents and other methods for solubilization of membrane proteins, together with interference problems in protein determination, have been fully discussed in relation to immunochemical analysis (15).

No definitive guidelines as to the use of detergents in gels and buffer compartments are possible. Incorporation of nonionic detergents into agarose may affect antigen mobility (92), the surface area delineated by immunoprecipitates (93,94), the morphology and sharpness of precipitates (93–96), and immunoprecipitation of lipoproteins (97). The presence of nonionic detergents will, however, prevent nonspecific precipitation or aggregation (392) and may be important when studying multienzyme complexes.

Ionic detergents, particularly SDS and sodium deoxycholate, possess relatively high critical micellar concentrations, in contrast with nonionic detergents (91), and have achieved widespread use in the analysis of membrane proteins. However, problems associated with alterations in antigenicity have limited the use of SDS for immunochemical analysis. Indeed, most enzymes are inactivated and irreversibly denatured by SDS, particularly if they are membrane bound (98). Nevertheless, recent studies indicate that under certain conditions, considerable "normalization" of precipitation patterns is possible for proteins solubilized or fractionated in the presence of SDS (97,99,100). In a recent CIE study of the effect of SDS on *M. lysodeikticus* ATPase (329b), Owen et al. observed that concentrations of SDS as high as 0.1% could be tolerated in antigen preparations, providing gels contained 1% Triton X-100 (see Fig. 8).

General aspects of the storage and stability of antigen preparations have been examined (17,101–103). Freezing or repeated freezing and thawing, particularly of detergent-solubilized extracts, can lead to aggregation (15,27), and even to total inactivation of enzyme components, as has been observed for the esterases from rat liver microsomes (94). Storage at 5°C can lead to changes in the immunoprecipitation patterns obtained with complex detergent-solubilized antigens (15,93,95,99). Membrane-associated protease (104) or other degradative enzymes, such as those derived from contaminating leukocytes, may be involved in such changes (15,95,105). Nonionic detergents may not inactivate contaminating or autogenous hydrolytic enzymes. Indeed, proteolytic enzymes, such as papain and pepsin, appear to be stable in the presence of generally denaturing concentrations of SDS (106), and may even be activated by detergents (107). The addition of appropriate protease inhibitors during membrane preparation and/or solubilization has been recommended (15,34,35,98,99), and the dispensing of such preparations into about 50-μl aliquots prior to storage appears to be a convenient method of minimizing aggregation caused by repeated freeze-thawing.

3. IDENTIFICATION OF ENZYMES

The establishment of a reproducible reference immunoprecipitate pattern on quantitative immunoelectrophoresis of complex antigen mixtures creates *per se* the subsequent challenge of identifying individual components. Unraveling the identity of each immunoprecipitate within such complex patterns has only been achieved for human serum proteins because of the availability of individual purified components and monospecific antibodies (108). For novel systems, such as the cytoplasmic fractions of *Pseudomonas aeruginosa* (109) and *N. gonorrhoeae* (24), this is almost an impossible task. In such instances, identification of enzymes by immunochemical staining techniques may provide a useful starting point (Fig. 1).

3.1 Enzyme Staining

The interaction of enzymes with their respective antibodies has been the subject of several reviews (110–114). Inhibition of enzyme activity may be complete, partial, or absent, and, in a few exceptional cases, stimulation of activity can result. In most cases, residual activity persists, even in extreme antibody excess. Inhibition has been related to aggregate formation (114) and to substrate size (110,111), although it should be noted that not all enzymes that have small molecules for their substrates behave similarly. In other cases, for example, penicillinase, conformational changes due to enzyme-antibody interactions appear to be involved (113,114). The heterogeneity of antibodies elicited in response to an enzyme can vary greatly with mode of administration and with host response, so that individual antisera differ in their capacity to interact with various antigenic determinants on the homologous enzyme. The relative positioning of the catalytic site in relation to other antigenic sites on the enzyme is thus important (114).

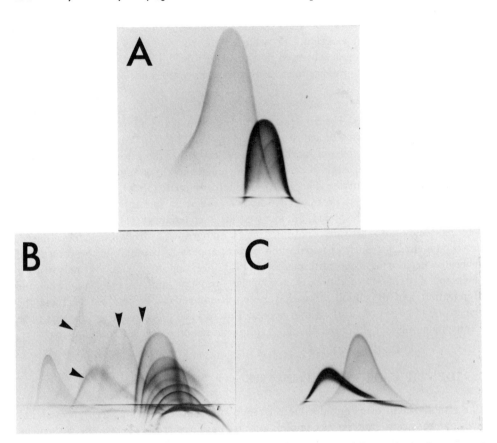

Fig. 1. Identification and enzymatic characterization of immunoprecipitates obtained on crossed immunoelectrophoresis of envelope antigens of *N. gonorrhoeae*. Envelopes prepared by freeze-pressing gonococci (Strain GCl) were washed four times prior to extraction with 4% (v/v) Triton X-100 (24). Aliquots of this extract (26.5 μg) were analyzed under identical conditions by CIE against purified antigonococcal envelope immunoglobulins. Gels were stained for lactate dehydrogenase activity (A), with Coomassie brilliant blue (B), and for NADH dehydrogenase activity (C). Immunoprecipitates that correspond to those that possess enzymatic activities are shown by arrows in B.

Recent evidence suggests some important differences between antigen-antibody immunoprecipitates in tube precipitin reactions and immunoprecipitation lines in quantitative immunoelectrophoresis experiments (2). At the zone of equivalence in the quantitative precipitin reaction, precipitation of antibody is maximal (115). Cann (116,117) has developed a phenomenologic theory for quantitative immunoelectrophoresis based on numerical solution of a set of mass transport equations to predict the developmental behavior of immunoprecipitates in RIE and CIE. The calculated evolution of the theoretical rocket immunoprecipitates mirrored most aspects of experimentally observed behavior. Initially, in quantitative immunoelectrophoresis, antigen molecules are in excess. Binding of antibodies to antigen leads to soluble immunocomplexes to which further amounts of antibody bind on migration, yielding eventually an insoluble precipitate. Once formed, such immunoprecipitates are usually unaffected by continued electrophoresis or electrophoretic traverse of antigens or antibodies (2,5). Systems that

involve horse antibodies may be exceptional due to the greater solubility of immunocomplexes in antigen or antibody excess. Harboe and Ingild (19) have reported that the
amount of antibody that produces immunoprecipitation in quantitative immunoelectrophoresis is some two- to 10-fold less than the "equivalent amount of antibody"
required in quantitative precipitin tests, the variation being dependent on the molecular
weight of individual antigens. Thus, enzyme-containing immunoprecipitates in quantitative immunoelectrophoresis may retain catalytic potential due partly to relative
antigen excess. This may explain in part why *M. lysodeikticus* ATPase can be revealed
by zymogram techniques following CIE of membrane extracts (8), even though *in vitro*
experiments indicate noncompetitive inhibition of the enzyme by antibody (see Ref.
118). Furthermore, an antienzyme antibody population consists of species that differ in
inhibitory capacity and specificity (see Ref. 114 for review). Data on the relative roles of
these different antibodies in immunoprecipitation in quantitative immunoelectrophoresis
are lacking.

Enzymatic staining of immunoprecipitates has been reviewed in theory and practice
by Uriel (119–121). More recent references to its application in quantitative
immunoelectrophoresis are given by Verbruggen (5) and Brogren and Bøg-Hansen (84).
Most substrates used for staining are synthetic low-molecular-weight compounds that
penetrate readily into immunoprecipitates. Chromogenic substrates yield colored insoluble products or poorly soluble products that are readily coupled to a diazonium salt to
yield a colored insoluble complex. Nonchromogenic substrates are linked to secondary
chromogenic systems that usually involve electron transfer to yield reduced insoluble
colored products. In this way, NADH dehydrogenase and lactate dehydrogenase have
been identified in the immunoprecipitation pattern obtained for Triton X-100 extracts
of *N. gonorrhoeae* envelopes (Fig. 1). Fluorescent substrates, such as umbelliferone derivatives, when hydrolyzed, deposit insoluble products on the immunoprecipitates that
then fluoresce in ultraviolet light. Hydrolases with high-molecular-weight substrates
have been detected by so-called print techniques (121–123) or by the use of agarose over
layers that contain substrate (124). It is also possible to detect biosynthetic activities by
the use of radioactively labeled precursors, if the end product is a polymer that remains
in situ or is enzyme-associated, for example, polynucleotide phosphorylase (121). In
other instances, enzyme staining involves a complex cascade reaction, such as that
utilized for the detection of enolase (125). This reaction is illustrated in Fig. 2.

Intrinsic properties of certain enzymes other than substrate specificity, for example,
binding properties, may be used to identify immunoprecipitates. Such "stains" may be
useful in the absence of suitable histochemical methods or when antibody completely
inhibits enzymatic activity. In quantitative immunoelectrophoresis, binding affinity has
been used to identify the [^{14}C]epinephrine receptors of rat liver plasma membranes
(126). Such an approach is now being applied in this laboratory for the identification of
the [^{14}C]penicillin-binding components of the D-alanine carboxypeptidase-transpepti-
dase system of a variety of bacterial species (127–129).

Many enzymes are metalloproteins that have prosthetic groups that contain metal
ions or that bind divalent cations (130). This binding was one of the criteria used by
Bock and Dissing (125) for the identification of enolase, that is, binding of ^{54}Mn^{2+} and
^{59}Fe^{2+}. Differences in the calcium-binding properties of prothrombin and of an
abnormal prothrombin induced by vitamin K deficiency and by dicoumarol administration have allowed their differential characterization by CIE (131,132). The presence of
phospholipids in enzyme-active immunoprecipitates has also been revealed by autora-

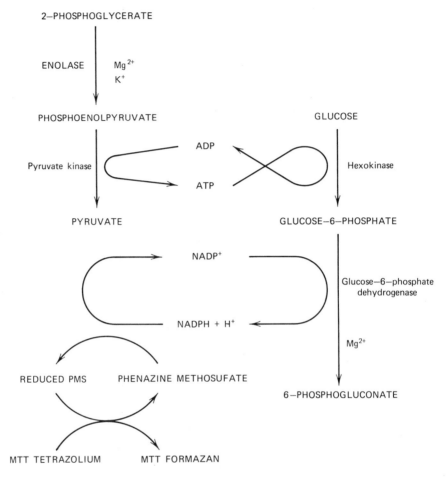

Fig. 2. Flow diagram of cascade reactions for staining for enolase. The incubation mixture contains substrate, two cofactors, three enzymes, MgSO$_4$, and KCl in Tris buffer. A blue to purple formazan is deposited on immunoprecipitates that possess enolase activity. The method also stains for adenylate kinase activity (see Ref. 125).

diographic techniques and by phospholipase C treatment (133). Reviews by Axelsen (14) and by Verbruggen (5) should be consulted for further examples of the use of radioactive tracer methods.

As with *in vitro* assays, enzyme inhibitors, such as fluoride, can be used in conjunction with quantitative immunoelectrophoresis methods to confirm the identity of an enzyme, such as enolase (125). Differentiation can also be achieved with inhibitors when two or more immunoprecipitates are identified with specific enzymatic activities. For example, diisopropylfluorophosphate, *p*-chloromercuribenzoate, and eserine can be used to differentiate esterases (15,84,96,133–135), and tartrate can be utilized to differentiate acid phosphatases (43,58,133). *p*-Chloromercuribenzoate has also been used to selectively inhibit NADH dehydrogenase and to differentiate between certain [^{14}C]epinephrine-binding proteins (126,133). In addition, differentiation is possible on the basis of substrate specificity, for example, α- and β-naphthylacetate for carboxylic ester hydrolases. An interesting variation of this principle has recently been used by Owen and Salton to distinguish between two membrane-associated NADH dehydrogenases from *M. lysodeikticus* (135a). A series of synthetic terminal electron acceptors suitable for detection of dehydrogenase activity and differing in electrode potentials are currently available from commercial sources. Tetranitroblue tetrazolium is the most electropositive and sensitive of those presently available. Clear distinction between the two NADH dehydrogenases was possible on the basis of their relative abilities to reduce these various tetrazolium salts (see Fig. 3). Only one dehydrogenase was able to reduce the highly electronegative acceptor tetrazolium red, whereas both enzymes utilized tetranitroblue tetrazolium readily. The presence of two distinct NADH dehydrogenases, as indicated above from differences in tetrazolium staining, was later confirmed by chromatographic studies (see Fig. 6).

The use of artificial substrates poses problems for enzyme detection and raises many questions. For example, carboxylic ester hydrolases, usually termed nonspecific esterases, comprise a broad group of enzymes of poorly defined specificity, the physiologic significance of which is unclear, with the exception of acetylcholinesterase. Enzymatic multiplicity has been demonstrated for esterases by crossed immunoelectrophoresis, especially with detergent-solubilized antigens from rat liver microsomes and plasma membranes (see Table 2). Such immunoprecipitates differed considerably in staining intensity and speed of development. Interpretation of faintly staining immunoprecipitates requires caution, because nonspecific coprecipitation of enzyme

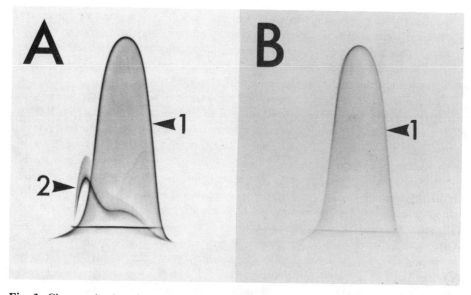

Fig. 3. Characterization of two discrete NADH dehydrogenase isoenzymes in membrane fractions of *M. lysodeikticus* by use of zymogram techniques. Membrane protein (57 μg) solubilized in 4% (v/v) Triton X-100 was subjected to CIE against antimembrane serum. Immunoplates were strained for NADH dehydrogenase (8) by use of tetranitroblue tetrazolium (A) or tetrazolium red (B) as electron acceptors. Unlike isoenzyme 1, isoenzyme 2 is incapable of reducing tetrazolium red (B). Similar phenomena were observed when other tetrazolium salts were used that have redox potentials more electronegative than MTT tetrazolium (e.g., neotetrazolium chloride and tetrazolium blue) (135a).

molecules with other proteins and their antibodies has been shown (15,84). In addition, some immunoprecipitates may possess nonspecific affinity for lipophilic substrates or coupling salts. Brogren and Bøg-Hansen (84) have recently reviewed criteria for enzyme characterization in immunoplates and the nature of possible artifacts.

Substrate specificities can also be difficult to define with histochemical stains, because qualitative assays fail to discriminate between the different affinities that apparently similar enzymes may have for a variety of substrates. The use of synthetic low-molecular-weight substrates of unknown physiologic significance, such as the L-glutamyl or L-leucyl-β-naphthyl derivatives used for detecting amidase activities (133,136), may increase the risk of falsely classifying them as similar enzymes, while quantitative assay of the same enzymes with their natural substrates, should they be known, might reveal differences in specificity. Other artifacts may arise from impurities in substrates. For example, the presence of ADP in purified ATP preparations may reveal both poly-nucleotide phosphorylase and ATPase (121,137). It is apparent from the above discussion that care should be excercised in interpreting multiple enzyme-active immunoprecipitates revealed by CIE as being indicative of multiple enzyme forms (138). Several procedures designed to confirm such an interpretation have been suggested by Brogren and Bøg-Hansen (84).

3.2   Alternative Methods of Enzyme Identification

The methods outlined below are particularly useful in identification when enzyme staining or binding properties cannot be utilized. The choice of any particular approach is usually dictated by the availability of the materials involved. For standard antigen-antibody systems, such as human serum, the principal antigens may be recognized simply by the appearance and pattern of immunoprecipitates and by the relative mobility of antigens (55,108). With other systems, identification of individual immunoprecipitates is possible if defined antigens or antibodies are available.

Because the surface area of an immunoprecipitate is dependent on the antigen/antibody ratio, coelectrophoresis of purified antigen with crude antigen samples will cause a specific increase in the surface area of one immunoprecipitate compared to a reference control immunoplate (8,55,108). Such an approach, which has been termed immunoenhancement (55), has been used to identify renal brush border alanine aminopeptidase in human urine (59). The same principle can be used for comparison of complex antigen mixtures to determine sharing of common components (55). When purified radioactively labeled antigens of high specific activity are available, identification is made by comparison of a stained pattern with an autoradiograph (139).

Other techniques that allow comparison of antigen preparations and establishment or interpretation of reactions of identity, nonidentity, and partial identity have been comprehensively detailed by several authors (16,27,140–143). Fused rocket immunoelectrophoresis (144), in which samples in adjacent wells are allowed to diffuse in antibody-free agarose prior to RIE, has been used to compare nucleoside di- and triphosphatases (NDPase and NTPase) and leucine aminopeptidase activities in detergent extracts of rat liver plasma membranes (126,145; see also Fig. 5). Rocket-line immunoelectrophoresis (RLIE), which is a combination of RIE and line immunoelectrophoresis [i.e., electrophoresis of an antigen from a trough or intermediate gel into a gel that contains antibody (146,147)] can be used to determine the degree of cross reaction between analogous proteins from different animal species (141). Tandem crossed immunoelectrophoresis (148), which may be regarded as a two-dimensional version of fused RIE, involves crossed immunoelectrophoresis of two antigen samples simultaneously from different antigen wells placed in such a way that diffusion for a period prior to electrophoresis will permit the observation of a reaction of identity in the form of a fused or double precipitin peak. This technique has been used to identify alanine aminopeptidase in the urine of a patient with acute tabular necrosis (149) and in studies on several peptidase systems (150–152). Crossed-line immunoelectrophoresis (CLIE) is a combination of CIE and line immunoelectrophoresis (153). Lactoperoxidase in bovine whey proteins (154) and glycophorin in solubilized human erythrocyte membranes (35,155) have been identified by this technique. The method also provides a means of demonstrating possible immunochemical relationships between electrophoretically separable, multiple forms of an enzyme, as demonstrated for enolase-active immunoprecipitates (125; see also Refs. 35 and 155).

The "absorption *in situ*" modification of CLIE (142,153) permits absorption of antiserum in the antibody gel during immunoelectrophoresis. An excess of complex antigen is used in the gel strip for line immunoelectrophoresis. Because of antigen excess, corresponding antibodies are removed by electrophoresis from the immunoplate as soluble immune complexes. This method has found application in the identification of species- or phase-specific antigens and of cross-reacting antigens and antibodies in microbial systems (22,23,28). Distinction between milk-specific and serum proteins in bovine whey has also been attempted with this procedure (154).

Several approaches are possible for the identification of specific immunoprecipitates when monospecific antisera are available. Addition of monospecific antiserum to a reference antiserum decreases the surface area of the appropriate immunoprecipitate compared to a control immunoplate, that is, immunosuppression (108,156). However, intermediate gel techniques with monospecific antibody allow easier evaluation and identification of specific components in complex reference immunoprecipitation patterns (157). This approach is illustrated in Fig. 6 for *M. lysodeikticus* ATPase. Rat brain enolase (125) and human serum cholinesterase (29) have been identified in a similar way. This technique has great potential for comparison of the antigenic diversity of enzymes within or between tissues.

Recently, a new technique for immunoprecipitate identification, termed line-absorption immunoelectrophoresis, has been introduced (100,158,159). Several wells that contain monospecific antisera are placed between the gel that contains the antigen and the gel that contains the polyvalent antiserum. During electrophoresis, antigens are absorbed by these monospecific antisera, resulting in dips in the line immunoprecipitates ahead of the wells in question. The technique permits identification of many proteins simultaneously with economical use of a battery of monospecific antisera.

Some other interesting modifications of CIE and CLIE with intermediate gels merit brief mention: (i) use of a series of antibody-containing gels of increasing titer (110,158; compare with Ref. 26); (ii) incorporation of each homologous antibody preparation into gels when comparing antigen preparations by tandem crossed immunoelectrophoresis (18,65); (iii) use of several intermediate gels with different antibody preparations (51,147); (iv) use of a split gel technique for the simultaneous analysis of a single antigen sample against two antibody preparations (393); (v) micromethodology for conserving antisera (4,161). The micromethod described by Weeke (4) has proved of exceptional value in the authors' hands. Complex antigen preparations, such as bacterial cytoplasm, can routinely be screened on the 5 × 5 cm glass plates suggested, and more than 70 immunoprecipitates have been readily resolved. The antibody volumes required vary but are of the order of 500 μl of serum per plate.

Purification of individual antigens for the production of monospecific antisera can be circumvented by immunizing animals with single immunoprecipitates excised from immunoplates. The higher resolving power of CIE or line immunoelectrophoresis allows easier identification and removal of appropriate immunoprecipitates on a preparative scale (86). This principle has been employed for the production of monospecific diagnostic antisera (162,163) and monospecific antibodies to phytohemagglutinin (164).

4. MODIFICATIONS OF BASIC METHODS AND COMBINATION TECHNIQUES

4.1 (Immuno)affinoelectrophoresis

Quantitative (immuno)affinoelectrophoresis involves electrophoresis of antigens in gels that contain free or immobilized lectins with or without subsequent electrophoresis into gels that contain antibody. This technique has opened up new potentialities for the biospecific identification of enzymes. Rocket affinoelectrophoresis denotes electrophoresis of antigen directly into gels that contain lectins and produces conical precipitates that can be employed to characterize enzymes (145,165,166) and to quantitate carbohydrate (167). Crossed immunoaffinoelectrophoresis (168) is a two-dimensional variant of CIE in which an intermediate gel that contains lectin is interposed between the gel strip that bears electrophoretically separated antigens and the reference gel that contains antibody. Binding to the lectin usually leads to retardation of glycoprotein in the gel that possesses lectin with loss or depression of corresponding immunoprecipitates in the reference antibody gel. Line-absorption immunoaffinoelectrophoresis is a recent modification that should allow simultaneous screening of antigens against several lectins (159).

To date, concanavalin A has been the most widely used lectin, although studies on *Ulex europeus* agglutinin, with specificity for carbohydrates that contain fucose (169), pokeweed mitogen (175), and *Ricinus* agglutinin, which has specificity for carbohydrates that contain nonreducing galactose residues (170), have recently been reported. In addition, soybean agglutinin, wheat germ agglutinin, and *Limulus polyphemus* agglutinin can be employed in intermediate gels (170a). Immunoaffinoelectrophoresis has recently been extended to the use of aluminum hydroxide in the intermediate gel. Human serum proteins were shown to vary widely in the extent of their absorption by this affinity absorbant. Because absorption appeared to be independent of net charge, molecular weight, or carbohydrate content of the proteins, it was concluded that the number of carboxyl guanidinium and carboxyl ϵ-amino groups might determine binding efficiencies (171). Quantitative immunoaffinoelectrophoresis has yielded data on concanavalin A-reactive enzyme antigens of rat liver plasma membranes (145), human urinary carboxylic ester hydrolases (135), and barley seed enzymes (166; see Refs. 8,92, and 172 for use of concanavalin A in other systems).

Bøg-Hansen and Brogren (173) have recently suggested that immobilized and free concanavalin A can be used to distinguish between glycoproteins with one or more binding sites for the lectin. Human urinary acid phosphatase was retarded by Sepharose-immobilized concanavalin A in an intermediate gel but not by free lectin, whereas human serum cholinesterase was retarded in both intermediate gels, indicating one and two or more concanavalin A-binding sites, respectively. However, broad generalization of this principle to all lectins in crossed immunoaffinoelectrophoresis awaits further investigation. It should be noted also that Harboe et al. (170) indicated that not all proteins that interact with *Ricinus* agglutinin form precipitates in agarose gels that contain terminal galactose groups. These residues compete with similar residues on glycoproteins for the lectin-binding sites. Thus, polyclonal IgG will precipitate with *Ricinus* agglutinin in phosphate-buffered saline but not in agarose gel (170).

If rocket affinoelectrophoresis is substituted for electrophoresis in agarose alone during the first dimension of CIE, heterogeneity based on qualitative and quantitative differences in carbohydrate composition of glycoproteins can be detected (169).

4.2   Combination Techniques

This category encompasses techniques that utilize different gel media, such as polyacrylamide, and/or different separation methods, for example, isoelectric focusing (crossed immunoelectrofocusing), polyacrylamide gel electrophoresis (disc crossed immunoelectrophoresis), and thin-layer gel filtration, instead of the first-dimensional zone electrophoresis in agarose gel (69,174–176). These methods often allow superior resolution of antigens.

Although starch gel was first employed as an alternative to agarose as a stabilizing medium in the first dimension (177,178), polyacrylamide gel has been favored by most investigators because of its lack of charged groups, the possibility of varying median pore radius of the gel, and the high resolution achieved (Table 1). A "moulding-in" technique has been most frequently used; that is, a polyacrylamide gel that contains separated antigens is moulded into antibody-containing or antibody-free agarose. To overcome difficulties caused by differences in electroendosmosis between the two gel media (179), attempts have been directed toward purifying or modifying the agarose (179–182) or toward adding to it methyl cellulose (183), linear polyacrylamide (181,184), or polyethylene oxide (181). In other studies, which often involved

Table 1. Combined Techniques for Quantitative Immunoelectrophoresis

First-dimensional separation	Medium	Technical variant	Reference
Electrophoresis	Polyacrylamide	Rod gel	160, 176
		Rod microgel	156,185
		Slab gel	183, 189
		Pore gradient gel	200
SDS electrophoresis	Polyacrylamide	Rod gel	100
		Slab gel	194
Isoelectric focusing	Polyacrylamide	Rod gel	380, 381
		Rod microgel	201, 382
		Slab gel	69–71, 179, 180, 188, 383, 384
	Cellulose acetate	Strip	385
	Agarose	Slab gel	182
	Agarose-acrylamide	Slab gel	181
Gel filtration	Dextran (Sephadex)	Thin-layer gel	199
	Agarose-acrylamide (Ultragel)	Thin-layer gel	199
Isotachophoresis	Polyacrylamide	Slab gel	403

microcapillary polyacrylamide gels or isoelectric focusing (see Table 1) or specially constructed apparatus (160,185), satisfactory results have been obtained with commercially available agarose. Mixed agarose-acrylamide gels have been tried with limited success in either the first dimension (184,186,187), the second dimension (188), or in both dimensions (181).

More recently, the distortions that arise from differences in electroendosmosis at moulded polyacrylamide-agarose junctions were circumvented by a "laying-on" procedure. In this method, polyacrylamide gel strips that contained antigens were applied onto the surface of an agarose gel that contained antibody (69–71,176,189), somewhat analogous to the original description of crossed immunoelectrophoresis by Laurell (9,10). This procedure is illustrated in Fig. 4, in which diphtheria toxin has been subjected to disc crossed immunoelectrophoresis.

Crossed immunoelectrofocusing has been applied to the study of salivary amylases (188), diphtheria toxin (70), and nonspecific esterases from detergent-solubilized rat liver microsomes (71). The latter application is illustrated in Fig. 5. Electrofocusing and crossed immunoelectrofocusing have only recently been performed in specially purified agarose with or without addition of linear polyacrylamide (181,182). Such media may be particularly useful in the future for the study of high-molecular-weight enzymes or multienzyme complexes. Moreover, combination of crossed immunoelectrofocusing with affinoelectrophoresis may give insight into the molecular basis of microheterogeneities seen in many glycoproteins (173,179–181,183).

Despite some anomalies with certain proteins, SDS-polyacrylamide gel electrophoresis has proved invaluable in protein chemistry, particularly in the study of membrane-bound proteins (98,190). The finding of an inverse relationship between the logarithm of the molecular weight of proteins and their electrophoretic migration rate extended its usefulness (191–193). However, the disruptive effects of free SDS on immunoprecipitation patterns, nonimmune precipitation reactions, and alterations in protein antigenicity have restricted immunochemical analysis of SDS-solubilized proteins (see Refs. 15,97,155,and 394). Recently, however, experimental conditions that alleviate such problems have been detailed (97,100,194). To facilitate SDS-polyacrylamide gel crossed immunoelectrophoresis, Converse and Papermaster (194) electrophoresed proteins separated on SDS-polyacrylamide gels through intermediate gels that contained Triton X-100. The nonionic and ionic detergents compete for the same binding sites on proteins, resulting in the formation of mixed micelles, thereby reducing the critical micellar concentration of the ionic detergent to the level of the nonionic detergent (91). SDS-solubilized proteins can also be transferred to nonionic detergent by addition of the nonionic detergent directly to the antigen sample (97,155). Removal of SDS by ultracentrifugation (195), by electrodialysis (196), and by the use of anion exchangers in the presence or absence of urea are other promising alternative approaches (197,198). Furthermore, Loft (100) successfully used extensive washing of SDS gels in excess buffer to reduce the concentration of monomeric SDS and salts (compare with Ref. 176).

Combination of SDS-polyacrylamide gel electrophoresis with line-absorption immunoelectrophoresis permits simultaneous determination of the identity and molecular weights of proteins (100,158). Thin-layer gel filtration immunoelectrophoresis may also find application in the determination of molecular weight of enzymes (199). Daussant and Skakoun (200) combined pore gradient electrophoresis with electrophoresis into antibody-containing gel to study purified β-amylases from barley and indicated its potential for investigation of the antigenic relationships between oligomeric

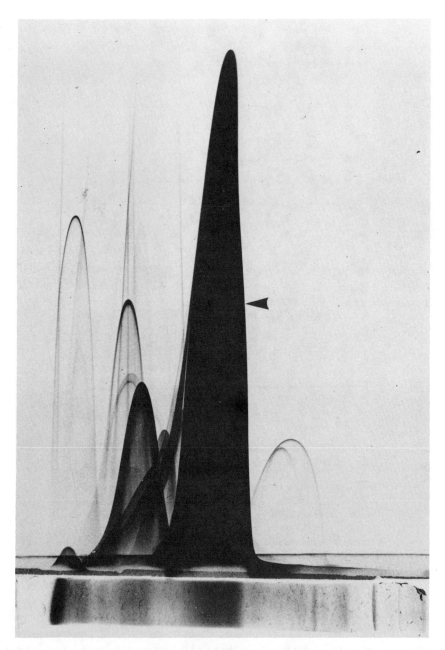

Fig. 4. Disc crossed immunoelectrophoresis of crude diphtheria toxin. Two hundred micrograms of concentrated culture filtrate were subjected to polyacrylamide disc gel electrophoresis, and the gels were subsequently sliced longitudinally and washed in distilled water to remove buffer ions (176). One half of the gel was utilized for electrophoresis into an agarose gel that contained antibody to diphtheria toxin (20 Lf/ml) by use of the "laying on" technique of Söderholm et al. (70). The remaining half polyacrylamide gel was fixed and stained for protein (inserted below immunoplate). Several immunoprecipitates are clearly resolved by this method; the major one (arrow) represents diphtheria toxin. Figure kindly supplied by J. Söderholm.

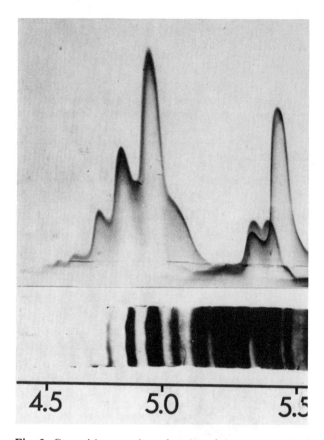

Fig. 5. Crossed immunoelectrofocusing of the esterases of rat liver microsomal membranes. Microsomal membranes were extracted for 30 min at 37°C in 50 mM Tris-HCl buffer (pH 8.5), and 300 µg of protein extract were subjected to crossed immunoelectrofocusing (70) against antiserum raised to rat liver microsomes. After electrophoresis, the immunoplate and control electrofocusing gel (inserted below immunoplate) were stained for nonspecific esterases with α-naphthylpropionate as the substrate. In the pH range shown, three immunoprecipitates that possessed distinct microheterogeneities are detected. Figure kindly supplied by K. Berzins.

forms of enzymes. Quantitative aspects of combined techniques have been reported by Jirka and Blanický (201) and Converse and Papermaster (194). In summary, this area of quantitative immunoelectrophoresis will probably continue to expand as investigators adapt techniques to solve specific problems.

5. FRACTIONATION AND PURIFICATION OF ENZYMES

Enzyme purification is often essential for a full understanding of enzyme structure and function. Various quantitative immunoelectrophoretic methods are available to monitor fractionation and purification. Analysis of selected or pooled fractions by CIE can be performed (62,92,126,181,202–204), although an accurate picture of the entire fractionation is obviously not possible. The use of CIE as a criterion of purity, however, is becoming more widespread (31,89,150–152,204–213). The enhanced resolving power makes it far superior to immunodiffusion or immunoelectrophoresis, but limitations with respect to the detection of basic proteins should be borne in mind.

For immunochemical analysis of column chromatographic and electrophoretic fractionation procedures, fused rocket immunoelectrophoresis (214) should be the method of choice, although its use in monitoring enzyme purification procedures (65,126,145) and in other fields is limited (65,139,140,157,164,168,214–218). It is rapid and considerably less tedious than other analytic methods, such as SDS-polyacrylamide gel electrophoresis or enzyme assay of individual fractions. Moreover, it not only displays the elution profile of the protein one desires to purify but also the distribution of other components, contaminants, or ballast protein in the material being fractionated, thereby permitting rational decisions regarding pooling of fractions consistent with maximizing both yield and purity. Its potential was illustrated in a recent report by Brogren and Svendsen concerning a multistep purification of human serum cholinesterase (219).

An example of fused rocket immunoelectrophoresis is shown in Fig. 6, where it has been used to monitor the fractionation of crude ATPase by gel filtration chromatography. An intermediate gel that contains antiserum specific for *M. lysodeikticus* ATPase has been introduced to aid in identification. It can be seen that the ATPase elution profiles monitored by fused rocket immunoelectrophoresis and by direct enzymatic assay are similar, suggesting that enzymatic activity is expressed to the same extent in all fractions. The protein profile also closely follows that of ATPase, except in the initial fractions, where contamination with other antigens clearly occurs. From an analysis of the fused rocket profile, it is possible to detect contamination due to other antigens in fractions 80–94 and 105–130. Therefore, for maximum yield of ATPase with minimal contamination, only fractions 95–104 would be pooled. Furthermore, if the fused rocket pattern contains at intervals wells with unfractionated material (see wells ii, Fig. 6), all antigens in the various fractions can be related to immunoprecipitates in the unfractionated material. In this way, it has been possible to show separation of the two NADH dehydrogenase isoenzymes on this column (antigens 4 and 5, Fig. 6; see also Fig. 3 for differentiation of these enzymes by staining procedures).

Modifications of CIE have been successfully used to predict the results of column chromatography (168). Bøg-Hansen showed that the results obtained by crossed immunoaffinoelectrophoresis with concanavalin A were completely analogous with those obtained by concanavalin A-Sepharose affinity chromatography of human serum proteins (168). Similar CIE experiments with affinity absorbents, such as aluminum hydroxide gels (171) and gels for molecular filtration (199), may also prove of value in screening for the purification method of choice for a particular enzyme.

6. ENZYME DISTRIBUTION

6.1 General Remarks

Quantitation of an enzyme is classically determined *in vitro* by assay of catalytic activity. This method has several inherent shortcomings. First, it measures only total activity and fails to distinguish between the presence of a single enzyme or several enzymes that have

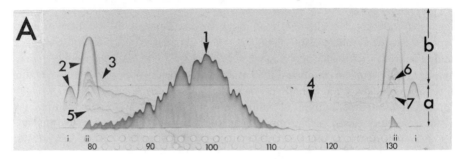

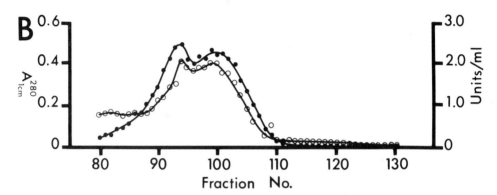

Fig. 6. Fused rocket immunoelectrophoresis of a crude ATPase preparation fractionated by gel filtration. ATPase was solubilized from *M. lysodeikticus* membranes by the butanol-shock procedure (272); 40 mg of protein were applied to an Ultragel ACA 22 column (135 × 2.5 cm) and were eluted with 50 mM Tris-HCl buffer (pH 7.5) that contained 10% (v/v) glycerol. Fractions (2.5 ml) were collected, and fractions 80–130 were monitored by fused rocket immunoelectrophoresis (A) and for protein (B) by extinction at 280 nm (O) and for ATPase by direct enzyme assay (●). One unit of enzyme activity is defined as the amount of enzyme that hydrolyzes 1 μmol of ATP/min. Gels marked a and b contain, respectively, monospecific antiserum to ATPase and antiserum to isolated membranes of *M. lysodeikticus*. The concentrations of antibody in gels a and b were adjusted to give similar titers of anti-ATPase. Wells marked i and ii contain 375 ng of purified mannan (263) and 22.4 μg of crude ATPase, respectively. Wells 80–130 contained 10 μl of eluted fractions. ATPase (antigen 1), identified by its inclusion in the intermediate gel as compared with a control (not shown), was distributed mainly in fractions 80–113. Fusion of the mannan immunoprecipitate (well i) with the strong immunoprecipitates for wells ii indicates the identity of the main antigen (no. 2) of the crude ATPase preparation. An antigen that shows partial identity with mannan can be seen in fractions 80–94 (antigen 3). Other antigens can also be detected in fractions 80–94 and 105–130. Two of these antigens (nos. 4 and 5) were identified by zymograms (8) as possessing NADH dehydrogenase activities. Some antigens, notably 6 and 7, were not detected in fractions 80–130. From Huberman et al. (329a).

similar substrate specificities. Second, it determines only active enzyme molecules and not those that have been denatured or inactivated by suppressors or inhibitors.

The environmental state of the enzyme, that is, whether it is soluble or membrane bound, and the conditions of assay can also greatly affect the determination of enzyme content (220). Many of these problems may be circumvented by the use of quantitative immunochemical techniques. For example, CIE allows the resolution and quantitation of isoenzymes. Thus, Verbruggen has recently demonstrated that crystalline subtilopeptidase A consists of two groups of antigenically distinct enzymes each of which can be resolved into four isofunctional forms (30).

As a general rule, quantitative immunochemical techniques, unlike direct enzyme assays, measure the total amount of enzyme present, that is, both catalytically active and inactive species. It should be remembered, however, that this situation only applies to enzymes whose antigenic integrity is preserved upon inactivation, inhibition, activation, or other modification. Thrombin, for example, does not react with antibodies to prothrombin from which it has been derived by proteolytic conversion (20). Thus, antiprothrombin cannot be used in the immunochemical estimation of this clotting factor. Chemically modified enzymes, such as carboxymethylated derivatives of lysozyme, papain, pepsin, and trypsin and performic acid-oxidized derivatives of bovine pancreatic ribonuclease, do not cross react with the native enzyme. Others, such as performic acid-oxidized staphylococcal nuclease, are fully cross reactive, however (113). It is apparent, therefore, that the immunologic relationships should be tested experimentally and not assumed (65,205,221). Thus, Verbruggen has shown by tandem CIE that subtilopeptidase A inhibited with diisopropylphosphofluoridate is antigenically identical to fully expressed enzyme (30). Similarly, the rocket height observed for fully active cholinesterase was shown to be identical to that obtained for a similar quantity of enzyme inhibited with eserine (84). In the latter investigation, electrophoretic separation of enzyme inhibitor during RIE could be ruled out, because the resultant immunoprecipitate could not be stained for cholinesterase activity (84).

The relationships between the results obtained from electroimmunoassays and direct enzyme assays have also been studied for plasminogen and for an extracellular protease (III) from *Staphylococcus aureus* (222). The latter enzyme was shown to require Ca^{2+} ions for maintenance of activity. Treatment with EDTA at 4°C produced a rapid and irreversible loss in its hydrolytic activity toward casein but, significantly, no decrease in peak height as monitored by RIE, which suggested that chelation of Ca^{2+} ions caused a conformational change in the protein in the vicinity of the active site. However, this change did not affect the immunologic specificity. In contrast, if incubation with EDTA was performed at 37°C, both enzyme inactivation and a decrease in rocket height were observed. Chromatography of the reaction products suggested that the enzyme had undergone autodigestion (222). Clearly, in this case and in others where enzyme inactivation or modification results in complete loss of determinants, for example, hydrolysis (222) or irreversible denaturation by ionic detergents (97), quantitative immunoelectrophoresis methods will only reflect the amount of native enzyme present. This is especially true for enzymes that possess subunit structure. Aging of alcohol dehydrogenase crystals, for example, has been shown to result not only in a loss of specific activity but also in a concomitant decrease in peak area after CIE. Furthermore, a corresponding increase was noted in the antigenically distinct peak that represented the alcohol dehydrogenase subunit (223). A very similar situation has recently been observed for *M. lysodeikticus* ATPase after treatment with low concentrations of ionic detergents (see Fig. 8). In these instances, the precipitin peak attributed to the native enzyme can only reflect the quantity of enzyme in the undissociated form.

If enzyme inhibition or inactivation causes partial loss of antigenicity, estimation of the total enzyme content by quantitative immunochemical methods becomes more difficult. In these instances, the relationships between the peak areas of active and inactive enzyme would have to be determined experimentally. An additional complication would arise if both forms of the enzyme were present in a sample to be assayed. Bock and Axelsen (16,224) have shown that, under certain conditions, RIE of two partially identical antigens can result in either one or two immunoprecipitates. In the former instance, the rocket height is lower than expected (presuming identity); in the latter case, both are higher than expected (presuming nonidentity), a situation that leads to underestimation or overestimation of content, respectively (16,224).

It is apparent, therefore, that only when the antigenic relationship between active and possible inactive enzyme forms has been established is it valid to draw conclusions about their distributions. For example, based on earlier immunochemical studies (222), Arvidson was able to conclude later that the low protease activity expressed in culture supernatants of *S. aureus* grown in Ca^{2+}-deficient medium reflected the absence of enzyme molecules and not the presence of enzymatically inactive protein (225).

Other phenomena that may lead to false estimation of enzyme distribution warrant brief mention. If the enzyme is an integral part of a multienzyme complex (see Ref. 133), or if it is complexed to another component of the antigen preparation, for example, human serum arylesterase with α_1-lipoprotein (84), the resultant peak area may not be the same as that obtained for an equivalent amount of free enzyme. Similarly, the relationship between peak height and enzyme content would not be expected to hold for isoenzymes, such as those of lactate dehydrogenase, which possess varying ratios of antigenically distinct subunits. Complex formation between enzyme antigens and serum components other than immunoglobulins may also cause artifacts (30).

If zymogram techniques are being used to monitor enzyme distribution, care should be taken to ensure that electrophoretic conditions do not cause an irreversible loss of catalytic activity. For example, it may be necessary to lower the operating pH conditions, as is the case for basic proteins (30,85), or to chemically modify the enzyme (48). The possibility that inhibition or inactivation may occur during initial fractionation procedures should always be borne in mind (96,145).

6.2   Enzyme Heterogeneity

The exceptional resolving power of CIE becomes apparent when one compares the number of antigens that can be detected by various immunochemical methods. For example, conventional immunoelectrophoresis studies by Blomberg and colleagues on rat liver microsomal membranes revealed the presence of six distinct esterases (134,137,226–228). Subsequent analysis by CIE resolved nine such enzymes (133,229), and analysis by crossed immunoelectrofocusing revealed as many as 20 (71; see Fig. 5). This same group has resolved 10 immunoprecipitates that possess nucleoside triphosphatase (NTPase) and/or nucleoside diphosphatase (NDPase) activities in the CIE reference patterns of both rat liver plasma membranes and rat liver microsomal membranes (133,229), as compared with six and three, respectively, by use of conventional immunoelectrophoresis (137,227,228,230,231; other examples can be found in Refs. 8 and 232–236).

CIE allows clear distinction between enzymes that are antigenically similar and those that are antigenically distinct or only partially related. This distinction is of particular importance for antigenically distinct isoenzymes that possess similar electrophoretic properties (8,65,150). CIE can, of course, detect enzyme heterogeneity based on differences in electrophoretic mobility (see Table 2), whereas crossed immunoaffinoelectrophoresis can detect heterogeneity based on either quantitative or qualitative differences in carbohydrate composition, for example, malt esterases (166; see also Refs. 169 and 172). Crossed immunoelectrofocusing has also been used to demonstrate a marked (micro)heterogeneity for salivary amylases (188), diphtheria toxin (70), and for the esterases of rat liver microsomal membranes (71; see Fig. 5).

Table 2. Literature Survey of Enzymes Analyzed by Quantitative Immunochemical Techniques

Enzyme	E.C. No.	Source	Reference[a]
Alcohol dehydrogenase	E.C. 1.1.1.1	*Saccharomyces cerevisiae*	223 (1-a, b)
Lactate dehydrogenase	E.C. 1.1.1.27	Human serum	290 (C-b)
		E. coli plasma membrane	(1-a, b)
Malate dehydrogenase	E.C. 1.1.1.37	*M. lysodeikticus* plasma membrane	8 (1-b)
		E. coli plasma membrane	(1-b)
Isocitrate dehydrogenase	E.C. 1.1.1.42	*M. lysodeikticus* cytoplasm	(1-b)
6-Phosphogluconate dehydrogenase	E.C. 1.1.1.43	*E. coli* plasma membrane	(1-b)
		E. coli cytoplasm	(1-b)
Glucose-6-phosphate dehydrogenase	E.C. 1.1.1.49	*N. gonorrhoeae* cytoplasm	(1-b)
Glycerol-3-phosphate dehydrogenase	E.C. 1.1.99.5	*E. coli* plasma membrane	(1-b)
Dihydroorotate dehydrogenase	E.C. 1.3.3.1	*E. coli* plasma membrane	(1-b)
Succinate dehydrogenase	E.C. 1.3.99.1	*M. lysodeikticus* plasma membrane	8 (1-b)
		E. coli plasma membrane	(1-b)
Alanine dehydrogenase	E.C. 1.4.1.1	*M. lysodeikticus* cytoplasm	(2-b)
Glutamate dehydrogenase	E.C. 1.4.1.4	*E. coli* plasma membrane	(2-b)
NADH dehydrogenase	E.C. 1.6.99.3	Rat liver plasma membrane	145 (b, c); 126, 133, 229 (10-b)
		Rat liver microsomal membrane	133, 229 (7-b)
		Plasma membrane of D23 hepatomas	136 (8-b, d)
		Plasma membrane of D33 hepatomas	136 (5-b, d)
		Microsomal membrane of D23 hepatomas	136 (6-b, d)
		Microsomal membrane of D33 hepatomas	136 (5-b, d)
		E. coli plasma membrane	(1 to 3-b)
		N. gonorrhoeae envelopes	24 (2-b)
		M. lysodeikticus plasma membrane	8 (2-b)
		M. lysodeikticus cytoplasm	(2-b)
Oxidase	E.C. 1.10.3.-	*Datura stramonium*	64 (2-b)
Catalase	E.C. 1.11.1.6	Rat liver plasma membrane	145 (1-b, d)
		M. lysodeikticus cytoplasm	(1-b)
Peroxidase	E.C. 1.11.1.7	Bovine whey	154 (1-b, d, e)
Ceruloplasmin	E.C. 1.14.18.1	Human serum	147, 153, 386 (1-e); 332 (1-a); 1 (1-a, b); 108 (1-a, f); see also Ref. 5.
		Bovine serum	158 (2-a)
Hexokinase	E.C. 2.7.1.1	Human tissue	211 (2-a, b)
Polynucleotide phosphorylase	E.C. 2.7.7.8	*M. lysodeikticus* cytoplasm	(1-b)
Nonspecific esterase	E.C. 3.1.1.-	Human cell line WI-38	43 (1-b, d)
		Human urine	135 (3-b)
		Rat liver plasma membrane	126, 133, 136, 145, 229 (1-b)
		Rat liver microsomal membrane	133, 136, 229 (9-b); 71 (20-b)

Table 2. (Continued)

Enzyme	E.C. No.	Source	Reference[a]
		Plasma membrane of D23 and D33 hepatomas	136 (1-b)
		Microsomal membrane of D23 and D33 hepatomas	136 (1-b)
		Rat liver lysosomal membrane	96 (8-b)
		Rat liver lysosomal content	96 (4-b)
		Barley malt	166 (1-b)
Aryl esterase	E.C. 3.1.1.2	Human serum	84 (1-b)
Cholinesterase	E.C. 3.1.1.7/8	Human erythrocyte membrane	36 (1-b, d); 15, 35 (1-b)
		Human serum	97, 108, 173, 219 (1-b); 29, 84 (1-a, b)
Alkaline phosphatase	E.C. 3.1.3.1	Rat chloroma Mia C51	221 (1-b)
Acid phosphatase	E.C. 3.1.3.2	Human cell line WI-38	43 (1-b)
		Human urine	58, 173 (1-b)
		Rat liver microsomal membrane	133, 229 (7-b)
		Microsomal membrane of D23 hepatomas	136 (5-b, d)
		Rat liver lysosomal membrane	96 (11-b)
		Barley malt	166 (1-b)
5′-Nucleotidase	E.C. 3.1.3.5	Rat liver lysosomal membrane	96 (6-b)
Phosphodiesterase	E.C. 3.1.4.1	Mouse liver plasma membrane	213 (1-b)
Phospholipase C (α-toxin)	E.C. 3.1.4.1	*Clostridium perfringens*	209 (1-f)
Aryl sulfatase	E.C. 3.1.6.1	Rat liver lysosomal membrane	96 (3-b)
		Rat liver lysosomal content	96 (1-b)
α-Amylase	E.C. 3.2.1.1	Human pancreas	205 (3-a, b)
		Human saliva	1 (1-a); 188 (2 to 4-a); 205 (5-a, b); 62 (1-b)
		Barley seed	65, 66 (2-a, b)
β-Amylase	E.C. 3.2.1.2	Barley seed	200 (6-b)
Lysozyme	E.C. 3.2.1.17	Human serum and urine	60 (1-a); 61 (1-a, d)
		Human sputum	62 (1-a)
β-Galactosidase	E.C. 3.2.1.23	Rat liver lysosomal content	96 (1-b)
N-acetyl-β-glucosaminidase	E.C. 3.2.1.30	Human cell line WI-38	43 (1-b)
β-Glucuronidase	E.C. 3.2.1.31	Human cell line WI-38	43 (1-b, d)
Protease	E.C. 3.4.-.-	Human leukocytes	294, 295 (C-g)
		Dog leukocytes	294, 387 (C-g)
		Dog plasma	299 (C-g)
		S. aureus	222, 225 (2-f)
Leucine aminopeptidase	E.C. 3.4.11.1	Rat liver plasma membrane	126, 133, 145, 229 (3-b)
		Rat liver microsomal membrane	133, 229 (2-b)
		Plasma membrane and microsomal membrane of D23 hepatomas	136 (2-b)
		Plasma membrane and microsomal membrane of D33 hepatomas	136 (1-b)
		Rat liver lysosomal membrane	96 (3-b)
		Rat liver lysosomal content	96 (2-b)

Table 2. (Continued)

Enzyme	E.C. No.	Source	Reference[a]
Alanine aminopeptidase	E.C. 3.4.11.-	Human brush border membrane	59, 149 (1-b)
Glutamyl aminopeptidase	E.C. 3.4.11.-	Plasma membrane of D33 hepatomas	136 (2-b, d)
		Microsomal membrane of D33 hepatomas	136 (4-b)
		Fetal rat liver microsomal membrane	136 (4-b)
Dipeptidase	E.C. 3.4.13.2	Pig small intestine	151 (1-f)
Prolidase	E.C. 3.4.13.9	Pig small intestine	152 (1-f)
Endopeptidase	E.C. 3.4.21.-	*Apis mellifica*	206 (2-b)
Chymotrypsin	E.C. 3.4.21.1.	Bovine pancreas	293, 298 (C-g)
		A. mellifica	206 (2-b)
		Micropolyspora faeni	388 (4-b, d); 389 (4-b)
		Aspergillus fumigatus	208 (5-b)
		E. coli plasma membrane	(1-b)
		E. coli outer membrane	(1-b)
Trypsin	E.C. 3.4.21.4	Bovine pancreas	32, 74, 297 (C-h); 289 (C-g)
		Dog pancreas	207 (F-b); 33 (C-b, h)
		A. mellifica	206 (2-b)
		Carcinus maenus	150 (4-b)
		Astacus leptodactylus	234, 241 (2-b)
		Astacus fluviatilis	234, 241 (5-a, b)
		Barley seed	65 (1-b)
(Pro)thrombin	E.C. 3.4.21.5	Human serum/plasma	131 (1-a); 289, 313 (2-a); 131, 311, 312 (F-a); 305 (C-a)
		Bovine plasma	132, 315, 316 (2-a); 315, 316 (F-a)
Plasmin(ogen)	E.C. 3.4.21.7	Human serum/plasma	301, 302 (C-g); 303, 304 (C-a); 304 (F-a)
Elastase	E.C. 3.4.21.11	Human granulocytes	212 (3-a); 292 (C-a); 296 (C-f); 295 (C-g)
Subtilopeptidase	E.C. 3.4.21.14	*Bacillus subtilis*	30 (8-f)
Collagenase	E.C. 3.4.24.3	Human granulocytes	210 (2-a); 292 (C-a)
Asparaginase	E.C. 3.5.1.1	*E. coli*	194 (1-a, d)
ATPase	E.C. 3.6.1.3	Rat liver plasma membrane	133, 229 (10-b)
		Rat liver microsomal membrane	133, 229 (7-b)
		Microsomal membrane of D23 hepatomas	136 (6-b)
		Microsomal membrane of D33 hepatomas	136 (5-b)
		E. coli plasma membrane	(1-b)
		M. lysodeikticus plasma membrane	8 (1-b)
NTPase/NDPase	E.C. 3.6.1.6	Rat liver plasma membrane	126, 133, 145, 229 (10-b)

Table 2. (Continued)

Enzyme	E.C. No.	Source	Reference[a]
		Rat liver microsomal membrane	133, 229 (7-b)
		Plasma membrane of D23 hepatomas	136 (8-b)
		Microsomal membrane of D23 hepatomas	136 (6-b)
		Plasma membrane and microsomal membrane of D33 hepatomas	136 (5-b)
Carbonic anhydrase	E.C. 4.2.1.1	Human erythrocytes	47, 48 (2-a)
Enolase	E.C. 4.2.1.11	Rat brain synaptosomes	125 (2-b)
Diphtheria toxin		*Corynebacterium diphtheriae*	70 (8 to 11-f); 89 (i)

[a] Numbers outside parentheses denote the relevant references. However, unpublished results from tha authors' laboratory concerning the plasma membranes of *E. coli* and cytoplasmic fractions of *M. lysodeikticus* and *N. gonorrhoeae* have not been given a reference number. The first number within parentheses refers to the number of isoenzymes and/or multiple forms detected, whereas C and F denote that either complex formation (C) or fragmentation (F) has been studied. Other letters in parentheses indicate the following: a, enzyme identified by use of monovalent antiserum; b, enzyme identified by zymograms; c, enzyme inactivated; d, no data shown; e, enzyme identified by line or crossed-line immunoelectrophoresis; f, enzyme purified; g, presence of enzyme in immuno-precipitate implied but not shown directly; h, enzyme localized by autoradiography; i, enzyme used in reversed rocket immunoelectrophoresis.

6.3 Localization of Enzymes

Since the development of quantitative immunoelectrophoresis, there has been an increasing recognition that these techniques can be of value in solving many diverse biologic problems. The following sections illustrate these problems, with special reference to comparative study of tissues, cells, membrane fractions, and so on. In some areas, enzymes *per se* have not been investigated but are included for completeness.

6.3.1 Tissue Specificity

CIE and CLIE have been used to identify brain-specific antigens in the rat (237,238). Analysis revealed that five of the 25 water-soluble rat brain antigens were organ specific (237) and were associated with the synaptosomal plasma membrane fraction (238). Two of these antigens were later demonstrated to possess enolase activity (125). Similarly, Bock has determined that only two of nine nonplasma cerebrospinal fluid antigens are unique to extracts of brain tissue (239).

The distinctive isoenzyme patterns of pancreatic and salivary amylases observed after CIE have allowed their detection in human urine (205) and their quantitation in both salivary and sputum proteins obtained from patients with chronic bronchitis (62). The origins of allergens in household dust (240) and in bovine whey proteins (154) have also been investigated by use of CIE and CLIE.

6.3.2 Determination of Taxonomic and Phylogenetic Relationships

Proteolytic enzymes have been used by Pfleiderer and colleagues as markers for determining phylogenetic relationships (150,206,234,241). Several endopeptidase isoenzymes were resolved by CIE in the gastric juices of both honeybees (206) and decapodes (150,234,241), and their immunologic relationships were determined (234,241). Thus, it was shown that three of the five endopeptidases of the crayfish *Astacus fluviatilis* reacted with antiserum raised to purified endopeptidases from the closely related species *Astacus leptodactylus*. However, no cross-reacting proteases were detected for the more distantly related decapode *Carcinus maenus* (234,241). Similarly, the immunologic relationships among α-amylases in varieties of barley have also been established (66).

CIE and CLIE have been found to be exceptionally useful tools in the study of different species and strains of microorganisms, for example, *Staphylococcus* (242), *Pseudomonas* (103,109,243), *Proteus* (233), *Mycobacterium* (27,235), .*Actinomyces* (244,245), *Chlamydia* (246), *Candida* (17), *Mycoplasma* (55,223), *Histoplasma* (247), and *Herpes simplex* virus (139,163,172,248,249). The cross reactivities of both *Pseudomonas* (22) and *Proteus* (233) with other bacterial genera have also been studied. It seems probable that many of the common antigens observed in these studies are enzymes, although, to date, none have been identified as such.

As part of her studies of brain-specific antigens, Bock has demonstrated by CIE and CLIE that only one of the 28 antigens detected for the rat brain was species-specific for the rat (237).

6.3.3   Enzyme Variations during Growth Cycle

The quantitative aspects of CIE allow changes in the concentrations of various enzyme antigens to be monitored simultaneously during the growth cycle, thus facilitating studies of cellular differentiation. For example, the production of two α-amylases in germinating barley seeds was followed immunochemically and shown to reach a maximum for both isoenzymes 7 days after germination. Thereafter, the quantity of both isoenzymes decreased (66). Because these results paralleled those obtained from direct enzyme assays, it was concluded that the changes in α-amylase activity reflected *de novo* synthesis and subsequent degradation of the isoenzymes (65,66). In a related study, the relative distributions of four aminopeptidases in shoots and rootlets of germinating barley have also been investigated (65). Antigens unique to both the yeast and mycelial phases of *Candida albicans* have been demonstrated by CIE (25) and by CLIE (28).

The levels of glucosaminadase, β-glucuronidase, and an esterase isoenzyme in the human diploid cell line WI-38 have been studied immunochemically in relation to senescence (43). Significantly, the levels of all of these enzymes, which are believed to be lysosomal in origin, increased as the cells aged. The levels of an acid phosphatase also became highly variable just prior to death of the culture. It is worth emphasizing that in this study (43), as in others (65,66), particular isoenzymes could be quantitated independently.

Neoplastic transformation has been studied by Raftell et al. in an immunologic comparison of the enzyme-active membrane antigens of 4-dimethylaminoazobenzene-induced rat hepatomas and adult and fetal rat liver (136) and by Daussant et al. in an investigation of the levels of auxin oxidase enzymes in plants after injury or induction of tumors by *Agrobacterium tumefaciens* (64). In the former study, only one of the esterase isoenzymes found in adult liver microsomes was detected in microsomes of hepatoma cells. Furthermore, both plasma and microsomal membranes from hepatoma cells had a reduced complement of antigens that possessed NTPase activity but did retain one that was common to fetal rat liver. Hepatoma microsomes also possessed four glutamine aminopeptidases with similar electrophoretic mobilities to those observed in microsomes of fetal liver. These findings suggested that the reappearance of fetal antigens in malignant cells may reflect general dedifferentiation (136).

A note of caution should be introduced here with regard to the interpretation of immunoprecipitation patterns in two different antigen-antibody systems. The presence of similar precipitation patterns after reciprocal testing with homologous and heterologous antisera does not necessarily indicate identity of antigens, because there are numerous instances of extremely low levels of immunogen promoting strong antibody responses (24,239) and of measurable levels of antigens, such as the purified mannan from *M. lysodeikticus* membranes (135a), producing little or no antibody response.

6.3.4   Cytoplasmic and Membrane-Associated Antigens

An important question in studies of biologic membranes is which components can be considered integral membrane constituents and which are true soluble or cytoplasmic contaminants. By use of conventional immunologic methods, Salton showed several years ago that membrane fractions obtained from cell lysates of *M. lysodeikticus* had to be washed a total of six times to free them from cytoplasmic contamination (251). More recent studies with CIE and CIE with intermediate gel and utilizing antibody preparations raised to washed plasma membranes and to the 200,000g soluble cytoplasmic fraction from *M. lysodeikticus* have shown that catalase, isocitrate dehydrogenase, and polynucleotide phosphorylase are true cytoplasmic markers (135a). Conversely, ATPase, succinate dehydrogenase, and malate dehydrogenase were demonstrated to be true membrane markers (135a), comparing favorably with results obtained by direct enzyme assay (252). However, two NADH dehydrogenase isoenzymes were distributed in both membrane and cytoplasmic fractions of *M. lysodeikticus* and were thus unsuitable as specific membrane markers. However, recent evidence suggests that one of the two soluble micrococcal NADH dehydrogenase isoenzymes may be unique to the cytoplasmic fraction (135a).

Generally, a membrane component is considered to be one whose concentration in the membrane is not decreased significantly by repeated washing in buffer. Quantitative immunochemical analysis of membrane fractions at various stages of the washing procedure should therefore distinguish between immunogens that are being progressively removed and those that are not (see Ref. 15). Smyth and coworkers have recently attempted to define conditions under which envelope fractions from *N. gonorrhoeae* can be considered free of cytoplasmic contamination (24). It was observed that the release of soluble cytoplasmic antigens from the envelopes was barely detectable by CIE after the fourth successive wash in dilute buffer. However, analysis of Triton X-100 extracts of six times washed envelopes by CIE by use of intermediate gel techniques revealed the presence of small amounts of material that reacted strongly with antiserum raised to the cytoplasmic fractions (24). It seems probable that this represents cytoplasmic material entrapped within membrane vesicles and thus may be an inherent contaminant of the system.

Berzins and coworkers have also effectively used CIE combined with zymogram techniques to detect major differences in the number of acid phosphatases, esterases, arylsulfatases, β-galactosidases, and N-acetyl-β-glucosaminidase enzymes between the lysosomal membrane of rat liver and the lysosomal content (96). It should be mentioned that the technique for obtaining lysosomes involved the use of the detergent Triton WR-1339; therefore, the distribution of antigens may not reflect the *in vivo* situation. An important point illustrated in this study, however, is that information gleaned from CIE-zymogram analysis should be correlated with information from direct enzyme assay. For example, it is known that 50% of the acid phosphatase activity is released from lysosomes upon disruption (253,254); yet, no immunoprecipitate with this activity could be demonstrated by zymograms for the lysosomal content. Because two enzyme-active bands were observed for this fraction on polyacrylamide gel electrophoresis, it appears probable that inactivation or complete inhibition of the enzyme had occurred on CIE (96).

One of the most widely used methods for determining the mode of association of components to membranes has been to monitor their distribution between soluble and sedimentable fractions after perturbation of the membrane, for example, by sonication or high- or low-ionic-strength buffers. Clearly, CIE offers a sensitive method for rapid and simultaneous analysis of the distribution of all immunogens after membrane perturbation. It has been successfully used in this respect in the analysis of the esterases of rat liver microsomal membranes (229), membrane antigens of rat brain (255) and mouse liver (203), and proteins of complement-treated human erythrocytes (256).

6.3.5   Species Specificity of Membrane Antigens

By use of a wide variety of CIE techniques, Bjerrum and colleagues have compared in detail the Berol EMU-043-soluble antigens of a variety of mammalian erythrocyte membranes (15,18,36). They have been able to show, for example, that many of the bovine erythrocyte membrane proteins show immunologic identity or partial identity with those of the human erythrocyte and that erythrocytes of many less related species contained two proteins, one of which was identified as spectrin, that were immunologically related to proteins of the human erythrocyte membrane. Partial identity among the main structural proteins of the bovine erythrocyte membrane and the bovine milk fat globule membrane has also been demonstrated with the intermediate gel technique (18).

6.3.6   Subcellular Specificity of Membrane Antigens

Into this category falls the detection of contamination of membrane preparations with antigens characteristic of other membrane fractions. Lysosomes, for example, are notoriously difficult organelles to fractionate to homogeneity; they are often contaminated with membrane fragments derived from the plasma membrane and from microsomes. Berzins et al. have tackled this problem in a comparative CIE study of the enzyme-active membrane antigens of the lysosome, microsome, and plasma membrane of the rat liver (96). Testing of microsomal and lysosomal membranes with both the homologous and the heterologous antisera indicated that these two membrane systems shared many enzyme-active antigens. However, no enzyme marker was found that, with certainty, could be used to monitor the lysosomal membrane fraction for microsomal or plasma membrane contamination. Identification of an immunoprecipitate that corresponded to the plasma membrane marker adenylcyclase (257), however, was not attempted. Enzyme-active antigen markers for the lysosomal membrane were found (96,133).

Comparative CIE studies of rat liver plasma and microsomal membranes have revealed major differences in the distribution of enzyme-active antigens between these two fractions (133,136,229). Only one of the nine esterases detected for the microsomal membrane was found in the plasma membrane (133); this esterase probably corresponds to a similar antigen detected in liver lysosomes (96) and microsomes of fetal liver and hepatomas (136). Other major differences between the two fractions were reflected in the number and degree of association of enzymes in immunoprecipitates (see Table 3). However, from these data, one cannot determine conclusively whether any of the aminopeptidases or nucleoside phosphatases found for both fractions were immunologically identical (133).

CIE has also been used to elucidate the function (or lack of one) for the intracytoplasmic membrane system of bacteria (258). Most bacteria that have been examined by electron microscopy appear to possess invaginations of the cytoplasmic membrane termed mesosomes (for review, see Ref. 259), the function of which is still a matter of speculation (260,261). The contents of the mesosomes, the mesosomal vesicles, have in many cases been isolated and their chemical and biochemical characteristics compared (261). In general, it appears from direct enzyme assays that mesosomal vesicles lack many respiratory chain enzymes and cytochromes associated with the plasma membrane. It can be argued, however, that because mesosomal vesicles are largely in the form of right-side-out membrane vesicles and tubules (262), substrate accessibility to enzyme complexes located on the inner membrane surface may be a determining factor that influences enzyme expression. Plasma membranes as generally isolated are, on the other hand, largely in the form of open sheets (262) and should not be affected in this way. Salton and Owen have attempted to resolve such problems by comparing the zymogram pattens of solubilized mesosomal and plasma membranes tested by quantitative immunoelectrophoresis (258). The enzymes NADH dehydrogenase, malate dehydrogenase, succinate dehydrogenase, and ATPase were markedly reduced or totally absent from mesosomal membranes of *M. lysodeikticus*, as compared with corresponding plasma membrane fractions. Mesosomal membranes were, however, enriched in a succinylated mannan (8). These results parallel almost exactly those obtained by direct chemical and enzyme assay of the fractions (262,263) and eliminate conclusively the suggestion that the mesosomal content represents the bacterial equivalent of the mitochondrial cristae (see Refs. 260 and 261).

Another problem of subcellular membrane differentiation recently tackled by CIE is that of the fate of synaptic vesicles after their fusion with the synaptic basement membrane (255,264). Tests of both synaptic vesicle and synaptic membrane antigens with their homologous or heterologous antisera did not rule out the theory of regeneration of synaptic vesicles by endocytosis (for review, see Ref. 265) but suggested that if this process did occur, it must be highly selective, because synaptic vesicles did not contain three major synaptic membrane antigens (264).

7. MEMBRANE ARCHITECTURE

A full understanding of the many functions attributed to the cell membrane requires a knowledge of membrane structure. Consequently, much effort has been devoted in recent years to probing the molecular arrangements of components within cell membranes. One method of approach is to determine the distribution of membrane components with respect to their expression on the inside face and the outer surface of the membrane. Various techniques are available for studies of this kind. Many rely on chemical modification of cell surface components by agents considered membrane impermeable (for reviews, see Refs. 258 and 266). Ferritin labeling can also be used to localize membrane antigens if monovalent antiserum is available (267). Another method that has been used extensively to assess the localization of membrane-associated enzymes has been that of comparative enzyme assay of parent cells (or protoplasts) and inside-out membrane vesicles (268,269).

A further method, which can be combined with the powerful resolving power of CIE, is that of immunoabsorption. Basically, this approach involves absorption of membrane antiserum with intact cells (or protoplasts in the case of organisms that possess cell walls) and with isolated membranes. If immunogens are expressed only on the outer membrane surface, antibodies directed toward them will be absorbed by both cells and membranes. Conversely, antibodies directed toward immunogens expressed solely on the inner membrane face will be absorbed only by isolated membranes. Several points should be borne in mind, however. Because two determinants per antigen molecule are necessary for the building of an antigen-antibody immunoprecipitate, it follows that an immunogen with only one determinant on the inner membrane face may not be detected in direct tests against antiserum absorbed with intact cells. (It may be possible to detect such a situation by use of the intermediate gel technique.) Identification of immunogens that have two or more determinants on the inner face of the membrane and one or more on the outer surface is also complex. It follows that the total antibody titer to such an immunogen should decrease on absorption with cells or protoplasts. Provided that the determinants expressed on the two faces are not identical, a selective removal of antibodies directed against only surface determinants should occur. Such a situation could theoretically be detected by comparison of absorbed and unabsorbed antisera in tests against isolated membranes (35,155). This, of course, assumes that the avidities of the different immunoglobulins for the antigen in question are similar. An immunogen that possesses more than one determinant that is not expressed on either surface (i.e., one that is embedded in the membrane) should be detected in tests that utilize antiserum absorbed with isolated membranes. However, it should be remembered that absorption with isolated membranes may also remove antibodies to antigens that are expressed along membrane fracture surfaces should these develop during membrane isolation (401).

Fukui et al. were the first to use the principle of immunoabsorption in a study of the localization of membrane antigens of *M. lysodeikticus* (7). Johansson and Hjertén later applied CIE to the analysis of absorbed sera and clearly showed that only one of the 20 membrane immunogens detected for *Acholeplasma laidlawii* was expressed on the mycoplasma surface (401). It is apparent from this study and subsequent reports (8,35) that CIE offers considerable advantages over conventional immunochemical procedures in the analysis of immunoabsorption experiments. Not only does it allow enhanced resolution of the membrane immunogens but also the quantitative nature of the technique enables the course of absorption to be followed with ease (8,401). Because the area under an immunopreicipitate in the CIE system is proportional to the antigen-antibody ratio (3), it follows that for a given amount of electrophoresed antigen, absorption of antibody will be reflected by an increase in the area under the immunoprecipitate. Conversely, antibodies directed against membrane-associated immunogens that are not expressed during absorption should be unaffected, resulting in immunoprecipitates of constant area. In this way, minor changes in the antibody titer to individual immunogens can be monitored. Figure 7 shows the effect of absorption of antimembrane

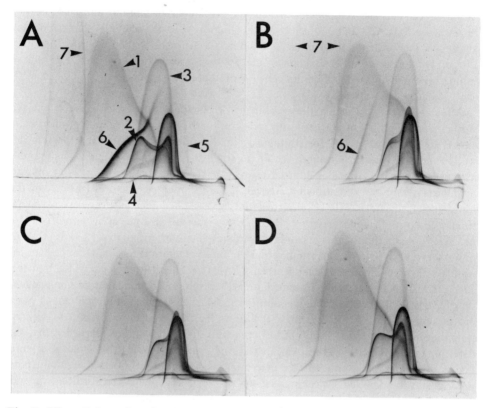

Fig. 7. Effect of absorption of antimembrane serum with protoplasts of *M. lysodeikticus* on the pattern obtained by CIE of a 4% (v/v) Triton X-100 extract of plasma membranes. Antimembrane serum was absorbed with 0 ml (A), 0.5 ml (B), 1.0 ml (C), and 3.0 ml (D) of washed protoplast suspension, and the immunoglobulin fraction was incorporated into agarose gels at equivalent concentrations. Membrane protein (57 μg) was analyzed by CIE in all instances. Areas under immunoprecipitates 6 and 7 (among others) increased on absorption (compare A with B) and were totally absent from patterns C and D. Immunoprecipitates 1–5 (among others) were unaffected by absorption. Data are taken from Ref. 8.

serum with protoplasts of *M. lysodeikticus* on the pattern obtained by CIE of a Triton X-100 extract of plasma membranes. It can be seen that antigens 1–5, among others, are unaffected by absorption, even by quantities of protoplasts three times that required to remove antibodies to antigen 6, a succinylated mannan (8), among others. It follows that the former group of antigens do not have any determinants expressed on the protoplast surface. By use of zymograms, it has been possible to show that these five precipitates possess the following enzyme activities: 1, succinate dehydrogenase; 2, ATPase; 3 and 4, NADH dehydrogenase; 5, malate dehydrogenase (8). Subsequent experiments performed with antisera absorbed with increasing amounts of isolated membranes indicated that all five identified enzymes were, indeed, localized on the inner membrane surface (135a). These results are in agreement with those obtained by Oppenheim and Salton in a ferritin labeling study of the localization of *M. lysodeikticus* ATPase (267).

An interesting observation in the absorption experiments with isolated membranes of *M. lysodeikticus* was that antibodies to membrane immunogens were not all absorbed at the same rate (135a). Similar phenomena have been observed by Gurd et al. in a CIE study of mouse liver plasma membranes (203). At present, it is uncertain whether these observations reflect the relative accessibilities of different membrane components, different concentrations and antigenicities of immunogens in the membrane, or both. Gurd et al. did find evidence to suggest that certain membrane antigens became more fully expressed after removal of other membrane components (203).

Absorption experiments performed with human erythrocyte membranes have shown the presence of four surface immunogens, two of which have been identified as cholinesterase and glycophorin (35,155). CIE with antiserum absorbed with intact erythrocytes in an intermediate gel failed to detect "inward feet" reactions, which suggested that neither component possessed determinants exposed on the inner membrane face (35,155). These results appear to correlate well with CIE studies of neuraminidase-treated human erythrocytes and of erythrocytes iodinated with ^{125}I by lactoperoxidase labeling (15). However, the latter methods were able to detect a unique surface glycoprotein (15) that was shown in CIE absorption experiments to lack determinants on the cell surface (35). This point illustrates one inherent disadvantage of the absorption method, namely, that only membrane immunogens that possess exposed regions in the form of determinants will be recognized as surface components.

8. MULTIENZYME COMPLEXES

A very interesting phenomenon has been observed by Blomberg and colleagues during immunochemical studies of rat liver plasma membranes (126,133,136,137,145,226,229–231), microsomal membranes (133,134,136,137,226–231,270), and lysosomal membranes (96). They detected in both immunoelectrophoresis (137,227,228, 230,231,270) and in CIE experiments (96,126,133,145,229) immunoprecipitates that possessed two or more distinct enzymatic activities. The authors termed these antigens "multienzyme complexes" and suggested that they reflect the organization of such components in the native membrane structure.

It will be seen from Table 3 that the plasma membrane reference pattern showed seven discrete immunoprecipitates that possessed four different enzymatic activities and that of the microsomal membrane showed 10 immunoprecipitates with three or more different enzymatic activities. Furthermore, autoradiography of immunoplates of both plasma and microsomal membranes labeled *in vivo* and *in vitro* with [^{14}C]choline and [^{14}C]ethanolamine suggested the presence of phospholipids in the immunoprecipitates that possess more than one enzymatic activity. This suggestion was confirmed by analysis of excised immunoprecipitates for [^{14}C]phosphatidylcholine and by phospholipase C digestion (133). More recent experiments that involved incubation of plasma membranes, plasma membrane extracts, and wet immunoplates with [^{14}C]epinephrine demonstrated that certain immunoprecipitates of the membrane reference pattern were capable of binding this hormone (126). To our knowledge, the results displayed in Table 3 are the only recorded instances of the detection by CIE of immunoprecipitates that possessed more than one enzymatic activity.

Because the identification of immunoprecipitates as multienzyme complexes could lead to the resolution of some of the complex functional associations that exist in the cell membranes, and because the interpretations of Blomberg et al. have met with some skepticism (84,271), a full discussion of these phenomena is felt justified. Primarily, the

Table 3. Occurrence of "multienzyme complexes" in membranes of rat liver[a]

Membrane fraction	No. of observed precipitates[b]	Enzyme activity detected							Other characteristics	
		NTP-ase	NDP-ase	Acid phosphatase	NADH dehydrogenase	Leucine aminopeptidase	Esterase	Aryl-Sulfatase	Phospholipids	Epinephrine receptors
Microsomal	7	+	+	+	+	−	−	nd[c]	+	nd
	3	−	+	−	−	−	−	nd	−	nd
	2	−	−	−	−	+	−	nd	−	nd
	9	−	−	−	−	−	+	nd	−	nd
Plasma	7	+	+	−	+	−	−	nd	+	+
	1	+	+	−	+	−	−	nd	+	−
	2	+	+	−	+	+	−	nd	+	+
	1	−	−	−	−	+	+	nd	−	−
	1	−	−	−	−	−	−	nd	−	−
Lysosomal	2	−	−	+	−	−	+	+	nd	+
	1	−	−	+	+	+	−	−	nd	nd
	1	−	−	+	−	−	+	−	nd	nd

[a] Data have been compiled from Refs. 96, 126, 133, and 229.

[b] Precipitates that possessed β-glucuronidase, β-galactosidase, or N-acetyl-β-glucosaminidase activities were also detected for the lysosomal membrane but did not appear to be associated with other enzyme activities (96) and thus have been omitted.

[c] Not determined.

question that has to be answered is whether multienzyme complexes are artifacts. Certainly, they do not appear to result from precipitation of enzyme-antibody complexes too close to one another to appear resolved. This hypothesis would require that many discrete enzymes possess not only identical charge heterogeneities but also equivalent antibody titers. Furthermore, manipulation of the antigen-antibody ratios did not resolve the various activities within immunoprecipitates (133).

Nor does the expression of two or more enzyme activities in one immunoprecipitate appear to result from a situation in which one enzyme possesses a wide substrate specificity. Blomberg and Raftell showed, by incubation of wet immunoplates with the enzyme inhibitors, that enzymatic activities in microsomal multienzyme complexes were dependent on at least three different catalytic sites (133). Similar results were obtained for multienzyme complexes of the plasma membranes (133). It seems probable, however, that the ability of most multienzyme complexes to hydrolyze both NTP and NDP may result from low substrate specificity (133,230; see also Ref. 213).

Neither does it seem probable from a study of precipitate morphology that the multiplicity of multienzyme complexes arises from proteolytic degradation caused by plasmin (34) or from endogenous membrane-associated proteases. However, a more serious argument is that multienzyme complexes arise from aggregation of components after membrane solubilization. This argument is strengthened somewhat by the fact that whereas Blomberg and colleagues routinely solubilized membrane fractions in 1% sodium deoxycholate plus 0.5% Lubrol W, they routinely performed CIE in detergent-free gels. Bjerrum and Lundahl have shown that CIE of detergent-solubilized human erythrocyte membranes in detergent-free gels results in diffuse immunoprecipitates, loss of resolution, and nonspecific precipitation near the application well (93,95). Indeed, most investigators who have studied membrane proteins have incorporated nonionic detergents into the electrophoresis gels. Nevertheless, Blomberg and colleagues maintain that their CIE reference patterns are highly reproducible (133) and that they are basically similar to the pattern obtained on inclusion of detergents into the agarose gels (133) or by an increase in the concentration of deoxycholate used for solubilization (229). These experiments, however, do not rule out the possibility that aggregation occurs after electrophoretic removal of deoxycholate. Indeed, CIE of deoxycholate extracts of gonococcal envelopes leads to nonspecific precipitation of protein, irrespective of whether the gels contain Triton X-100 (24).

If nonspecific association of membrane components does occur after solubilization, the process must be highly selective. Few, if any, of the immunoprecipitates showed lines of partial identity, as might be expected if the complexes arose by a random association of antigens after solubilization. Furthermore, the patterns were very reproducible with respect to the number of precipitates and to the enzyme activities that they expressed (96,126,133,145,229). Furthermore, phospholipids were only found associated with immunoprecipitates that displayed the properties of multienzyme complexes (133; see Table 3).

Blomberg and Raftell also attempted to determine whether enzyme polymorphism was due to interaction of enzyme molecules with detergent micelles by comparing the CIE patterns of Lubrol-deoxycholate membrane extracts with those of extracts solubilized by other agents, such as Triton X-100, papain, and phospholipase A or C (229). Of the latter, only treatment of microsomal membranes (and not plasma membranes) with dilute papain was shown to provide extracts that yielded immunoprecipitates that possessed multienzyme properties. Co-CIE of an ^{125}I-labeled papain extract with an unlabeled Lubrol-deoxycholate extract did, nevertheless, suggest identity (229). It is unfortunate perhaps that Raftell and Blomberg did not compare the enzyme patterns of the papain digest of [^{14}C]choline-labeled microsomal membranes with those of similar digests treated with phospholipases, especially since autoradiography (34) and phospholipase action (229) indicated that phospholipids appeared to be essential for the integrity of multienzyme complexes.

Close examination of the CIE immunoplates suggests that nonspecific staining or enzyme entrapment (84) could account for the identification of some, but by no means all, of the immunoprecipitates assessed as possessing certain enzyme activities. In these cases, regions of immunoprecipitates that intersected other stained immunoprecipitates are heavily reinforced. However, the fact that many of the immunoprecipitates showed complete staining over their full length and that some possessed the same enzyme activity but different substrate (133) or inhibitor (126,133) specificities speaks against this type of artifact occurring in all instances.

Attempts to fractionate the various multienzyme complexes of rat liver membranes either by affinity chromatography (145) or Sepharose 4B chromatography (126) were only partially successful. By the latter technique, antigens previously identified as multienzyme complexes of the plasma membrane were distributed over a wide molecular weight range (126). Significantly, fractions eluted in the high-molecular-weight range showed three NTP/NDPase-active precipitates with leucine aminopeptidase activities, whereas identical immunoprecipitates that stained for NTP-NDPase in lower-molecular-weight ranges did not possess aminopeptidase activities. A single immunoprecipitate with this activity was detected, but it did not appear to be associated with any other enzyme activity. Unfractionated material, on the other hand, gave a CIE pattern that showed two immunoprecipitates with both NTP/NDPase and leucine aminopeptidase activities, among others, and one that possessed aminopeptidase activity alone (see Table 3). These observations appear to support the thesis that the multiplicities of immunoprecipitates could in part reflect relative degrees of dissociation of molecular aggregates. Alternatively, the complexes could reflect genuine differences in the molecular associations that exist in the membrane and that are exhibited only by response to particular dissociation procedures. The results are not too dissimilar from those obtained in the detection of a Triton X-100-solubilized ATPase complex clearly different in electrophoretic properties from ATPase isolated by the butanol shock-wash procedure (272; compare Figs. 7 and 10). In this case, however, only ATPase enzyme activity has been detected in the Triton X-100-solubilized complex (8).

In summary, the evidence presented for the existence of "multienzyme complexes" in detergent-solubilized membranes is compelling but by no means conclusive. Certainly, there is ample precedent in the literature for the existence of multienzyme complexes both in soluble (for reviews, see Refs. 273 and 274, also Refs. 275–283) and in membrane-associated states (274,279,284–288). However, in many of these instances, the association of enzymes can be shown to reflect coupled biologic function, as, for example, in biosynthesis of fatty acids (274,395), tryptophan (274), and pyrimidine (275–278, 280, 281), in carbon dioxide fixation (282,283), and in electron transport phosphorylation (284–288). In contrast, the biologic significance of multienzyme complexes in rat liver membranes remains obscure. The purification of a discrete complex that possesses various enzyme activities that can be separated on removal of stabilizing forces (275,278,281–283) would certainly help to substantiate their existence. Near-neighbor mapping of membrane lipids and proteins (for review, see Ref. 258) may prove to be a more fruitful approach, however.

9. COMPLEX FORMATION

Complex formation, that is, the association of two or more discrete components to form a complex, is an integral part of most reaction processes. Complexes may be short-lived, as in many chemical reactions, or relatively stable species, as in most precipitation reactions. Detection of the complex is usually based on some unique property, such as molecular weight, absorption characteristics, and electrophoretic mobility. The ease with which CIE can detect heterogeneity makes it a method well suited to the study of complex formation, providing, of course, that the constituents are antigens and that a change in electrophoretic mobility results on complexing. Quantitation of the relative amounts of complex and free antigen is also possible (289), thus facilitating equilibration studies. CIE has been used most extensively in studies of complexing processes of clinical significance, such as the detection of soluble immune complexes (290,291,396). The formation of complexes between proteases and serum components, such as α_1-antitrypsin and α_1- and α_2-macroglobulins, after tissue damage has also been extensively investigated by this method. Normally, it is necessary to follow complex formation by use of antiserum directed against the serum component, because the latter may partially or completely mask determinants on the smaller enzyme molecules (289; see, however, Ref. 292). Nevertheless, the detection of both complexing antigens in one immunoprecipitate is necessary before complex formation can be assumed.

Often the species that complexes with the enzyme may inhibit enzyme activity, as observed for α_1-antitrypsin and α_2-macroglobulin (32,293–296), although the latter will generally allow the expression of enzyme activity if low-molecular-weight substrates are used (32,293). Thus, detection of enzymes by zymogram methods may often be unsatisfactory. Ohlsson has circumvented this problem by using ^{125}I-labeled enzymes to detect complexes formed *in vitro* between bovine trypsin and both dog α_1-antitrypsin and dog α-macroglobulins (31) and to measure the relative affinities of formation (32). Subsequent experiments performed with purified ^{125}I-labeled trypsin from the dog (207) demonstrated similar affinity ratios, suggesting that α-macroglobulins function as a defense mechanism against trypsin liberation in the dog. Similar experiments with both dog and human sera showed that ^{125}I-labeled chymotrypsin had greater relative affinity for α_1-antitrypsin than for α_2-macroglobulins (293). These observations were later confirmed by *in vivo* infusion experiments in the dog, for which the first signs of crisis were observed on complete complexing of serum α-macroglobulins and α_1-antitrypsin by trypsin (297) and chymotrypsin (298), respectively. The course of saturation of dog protease inhibitors has also been followed by CIE *in vitro* with leukocyte proteases (294) and *in vivo* during pancreatitis (299).

In a similar series of *in vitro* experiments, human leukocyte proteases (300) have been found to bind preferentially to human α_1-antitrypsin (294,295). In contrast, human granulocyte elastase (212) could be shown to form complexes with both α_1-antitrypsin and α_2-macroglobulins in inhibitor:enzyme stoichiometric ratios of 1:1 and 2:1, respectively (296). In a subsequent clinical study, both free and complexed human granulocyte collagenase (210) and elastase (212) were detected in samples of purulent sputum for which the protease-inhibiting capacity was fully saturated (292).

Grubb has also used CIE to demonstrate the presence of soluble IgG-lactate dehydrogenase immunocomplexes in the serum of a patient with a history of hyperthyroidism and generalized arteriosclerosis (290). Indeed, CIE and RIE can be used effectively to detect and quantitate immunoglobulins that do not form insoluble immune complexes (291).

CIE studies on rat serum have also revealed that a component called α_2-acute phase globulin, which appears in high concentrations in acute inflammatory responses, forms complexes with both ^{125}I-labeled plasmin (301) and ^{125}I-labeled trypsin (74) in an analogous manner to human α_2-macroglobulin.

Immunochemical techniques combined with gel filtration studies have been used to follow plasminogen activation by biologic activators and plasmin inhibition in plasma. Niléhn and Ganrot (302) demonstrated that plasmin was complexed to α_2-macroglobulin in human sera from patients undergoing intravenous infusion of streptokinase. Plasmin-α_2-macroglobulin complexes were also identified in human fibrinolytic postmortem plasma (303) and in urokinase-activated human plasma (304). Such complexes possessed benzoylarginine ethylesterase activity (303,304). None of the other plasma protease inhibitors tested, namely, α_1-antitrypsin, inter-α-inhibitor, antithrombin III, and C_1-esterase inhibitor, were found to complex with plasmin (302–304).

By use of a technical modification that involves incorporation of heparin into agarose during the first-dimensional electrophoresis, differences in electrophoretic mobility were demonstrated between antithrombin III in citrated plasma and serum (305). These changes were also shown to occur on spontaneous clotting of blood. Currently, the identity of the activated clotting factors complexed to antithrombin III is unknown.

Complex formation has also been demonstrated by CIE between subtilopeptidase A and serum components other than immunoglobulins (30) and between serum arylesterase and serum α_1-lipoprotein (84). It may also be possible to assess complex formation by excision of a particular precipitate and subsequent analysis of immune serum raised to it (86) or by introducing into the first-dimension gel absorbents, such as lectins, that display differential affinities for the components of the complex (169).

10.   FRAGMENTATION, MODIFICATION, AND DISSOCIATION OF ENZYMES

Fragmentation and dissociation of antigens can also be readily studied by quantitative immunoelectrophoresis, provided that the split products show different electrophoretic mobilities from the parent molecule and provided that their immunologic properties are not altered or destroyed. Quantitative data on the distribution of reaction products are possible as long as standardization can be achieved. However, in some instances, differences in the solubility of antigen-antibody complexes formed by small and large fragments may obviate the use of the relative peak heights or surface areas as direct measures of their relative concentrations, for example, prothrombin fragments (131) and fibrinogen split products (68).

Many animal proteases exist *in vivo* in the form of inactive precursors termed proenzymes or zymogens, for example, trypsin, chymotrypsin, plasmin, thrombin, pepsin, and carboxypeptidase. Conversion of zymogens to the active enzymes involves hydrolysis of peptide bonds with removal, in some cases, of a small peptide. For example, activation of trypsinogen to trypsin (306) results in the release of a hexapeptide from the amino terminus of the molecule. In other instances, a considerable portion of the zymogen is removed, as, for example, during activation of procarboxypeptidase A (307). The immunologic relationships between enzyme-proenzyme pairs have been studied to investigate conformational changes involved in activation (113,114). In some instances, antiproenzyme antiserum is extremely specific, reacting only with the homologous antigen and not with the active enzyme generated, for example, prothrombin (20) and procarboxypeptidase B (308). Other enzyme-proenzyme pairs appear to be immunologically identical, for example, trypsinogen-trypsin (309) and plasminogen-plasmin (310). Thus, CIE was used to follow direct proteolytic conversion of trypsinogen to trypsin (207).

The complex sequence of reactions that result in the formation of a blood clot involves several proteolytic enzymes that undergo conversion from an inactive to an active form. The only reaction extensively studied by CIE has been prothrombin activation (20,131,132,214,311–316). Activation of prothrombin led to the release of a derivative with higher mobility than prothrombin (131), which was shown to be the main nonthrombin fragment (311). This "profragment" could be resolved into two electrophoretically separable molecules that, in the presence of sodium citrate or Ca^{2+} and tissue extracts, were cleaved to a large fragment and two small fragments (converted profragments; Ref. 20). Antiprothrombin sera reacted with the profragment but not with thrombin, thrombin complexes in serum, or the smaller converted profragments (20).

Physiologic activation of prothrombin requires Ca^{2+}. Dicoumarol administration in both man and animals leads to the appearance of an abnormal prothrombin in sera identifiable by CIE (311,313,314). Normal prothrombin has three to four Ca^{2+}-binding sites that show cooperativity, whereas dicoumarol-induced prothrombin can only bind one Ca^{2+} ion per molecule and remains unchanged after physiologic activation of normal prothrombin. These differences in binding properties produce differences in electrophoretic mobility between the two prothrombin species (315–318).

Comparative biochemical and immunochemical studies of the amino terminal fragments produced by thrombin digestion of normal and dicoumarol-induced prothrombins revealed that although they were identical in molecular weight, amino acid composition, and carbohydrate content, the former bound Ca^{2+} and possessed Ca^{2+}-dependent antigenic determinants, whereas the latter did not (315,319). Because differences between the two prothrombins were confined to the amino terminal fragments (315,317), it appears that Ca^{2+} binding to this terminus of normal prothrombin involved a conformational change (319) essential for activation of prothrombin under physiologic conditions.

Thrombin catalyzes the conversion of fibrinogen to monomeric fibrin, which self-associates to form a blood clot. Degradation of fibrin deposits in blood vessels is the physiologic function of plasmin. Plasmin is generated from plasminogen by tissue activators, for example, urokinase, an endogenous activator found in kidneys and the urinary tract, or by exogenous activators produced by bacteria, for example, streptokinase from *Streptococcus pyogenes* and staphylokinase from *S. aureus*. Characterization of plasminogen and its activation by urokinase have recently been reviewed (320).

The sequence of events that occur after urokinase-mediated proenzyme-enzyme transformation has been studied in human plasma by CIE (304). At low urokinase concentrations, an enzymatically inactive component with increased anodal migration was detected. This intermediate could not be correlated with components that occurred in purified preparations of plasminogen or plasmin (320). Fibrin triggered conversion of this intermediate to plasmin, which was detected as a complex with α_2-macroglobulin. Similar processes occurred without fibrin at higher urokinase concentrations; the appearance of plasmin-α_2-macroglobulin complex was correlated with slow fibrinogen degradation. The production of plasmin appears to be catalyzed by the fibrin surface, with subsequent preferential degradation of fibrin and the formation of plasmin-α_2-macroglobulin complexes.

The interaction of plasmin with fibrinogen leads to the production of five to 10 split products, depending on the isolation technique (68). DEAE-cellulose chromatography revealed five fractions, identified as A, B, C, D, and E in order of elution (68). The main components (D and E) were of high molecular weight and possessed the fibrinogen-specific antigenic determinants. The other split products did not interact with antifibrinogen serum. Niléhn has characterized the D and E fractions by CIE (68). The D fraction was electrophoretically heterogeneous; it comprised nine separable components on prolonged electrophoresis, eight of which were immunochemically similar. By contrast, the E component was electrophoretically homogeneous. A sensitive method for the determination of proteolytic activity has been based on the immunochemical quantitation of these fibrinogen fragments (321). Breakdown products of fibrinogen after degradation with synovial tissue extracts from rheumatoid and nonrheumatoid arthritis patients were compared with those of plasmin by CIE (322). The latter degraded fibrinogen to D and E fragments, whereas the former produced incomplete degradation products, which suggested complex formation between rheumatoid synovial components and fibrinogen breakdown products. Fibrinogen break-down has also been monitored by CIE after intravenous administration of trypsin and chymotrypsin to dogs (297,298).

Alteration of CIE reference antigens on storage has been investigated, for example, in human serum (323–325), allergens (102,202), and detergent extracts of erythrocyte membranes (95). In the latter instance, the observed changes in precipitate morphology obtained with stored antigen were similar to those induced by pronase treatment (99). Antigens that contain sialic acid residues and their desialylated derivatives after neuraminidase treatment appear to be cross reactive qualitatively and quantitatively (15,68,92,326). The effects of controlled proteolytic degradation on human serum and urine proteins (99) and of fungal proteases on human serum proteins (327) have been followed by CIE, as has autodigestion of proteases (30,222,225).

11. ENZYME STRUCTURE

The potential of CIE for the analysis of enzyme subunit structure was first recognized by Linke and colleagues (223). Aging of crystals of yeast alcohol dehydrogenase for several months at 4°C was shown to cause a decrease in the area of the enzymatically active alcohol dehydrogenase peak and a concomitant increase in the area of a peak of slightly increased electrophoretic mobility. This peak, which was not enzymatically active and did not cross react with the parent molecule, was shown to correspond to the alcohol dehydrogenase subunit obtained from density gradient centrifugation experiments (223).

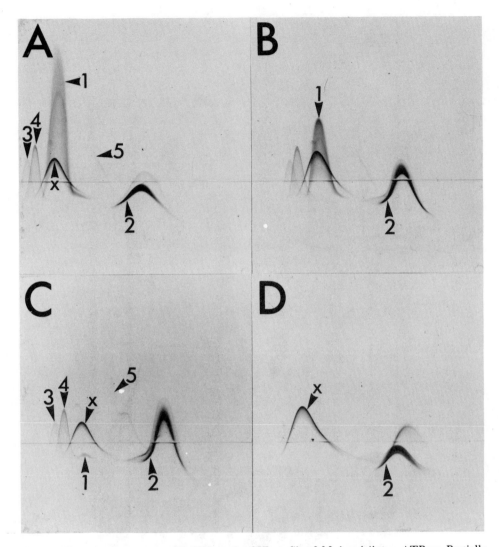

Fig. 8. Effect of pretreatment with SDS on the CIE profile of *M. lysodeikticus* ATPase. Partially purified ATPase (1.8 mg/ml) was pretreated with SDS at final concentrations of 0% (A), 0.05% (B), and 0.1% (C) (w/v), and free SDS was complexed by subsequent addition of Triton X-100 to a final concentration of 4% (v/v) (97). Thirteen micrograms of protein were analyzed by CIE against antiserum raised to an ATPase preparation that had been fully dissociated with urea, dithiothreitol, and iodoacetamide. Gels also contained antiserum raised to human transferrin. The native ATPase precipitin peak 1 decreased on addition of SDS (compare A, B, and C), and the area under precipitin complex 2 increased. Other minor contaminating components (nos. 3–5) remained virtually unaffected. D shows the CIE profile of ATPase extracted with 0.1% (w/v) SDS after electrophoresis in polyacrylamide gels and enzymatic localization (379). Only precipitin complex 2 is observed in D. Precipitate x in A–D represents the marker protein carbamylated transferrin (46). From Owen et al. (329b).

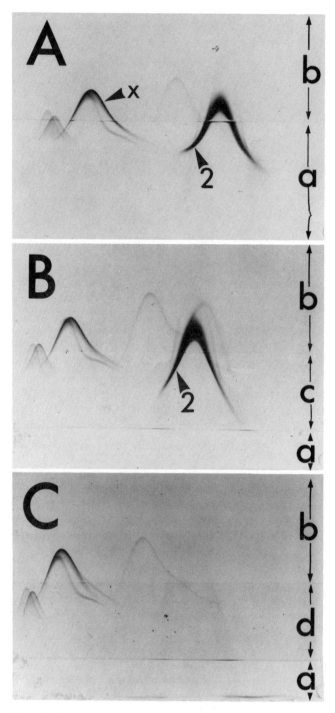

Fig. 9. Immunologic reactions of the subunits of *M. lysodeikticus* ATPase. Partially purified ATPase was dissociated with urea, dithiothreitol, and iodoacetamide, then dialyzed against distilled water, and finally made 4% (v/v) with respect to

It is fortuitous that, in the case mentioned above, antiserum raised to freshly isolated crystalline alcohol dehydrogenase produced antibodies to both the parent enzyme and to the immunologically distinct subunit (107). This does not appear to be the case with the membrane-associated ATPase from *M. lysodeikticus*. This enzyme, which can be shown by SDS-polyacrylamide gel electrophoresis to possess two major subunits (α and β) and two or three minor subunits (γ, δ, and ϵ; see Refs. 328 and 329) loses almost all of its ability to form precipitating immune complexes if dissociated into its substituent subunits (135a; compare with Ref. 7). Owen and Salton have recently raised antiserum to an ATPase preparation fully dissociated by treatment with urea, dithiothreitol, and iodoacetamide and have shown that this serum will not only form precipitates with some (or all) of the substituent subunits but will also react with the native enzyme. This phenomenon is illustrated in Fig. 8, which shows the effect of increasing concentrations of SDS on the CIE pattern of partially purified ATPase. The decrease in the height of peak 1, which can be shown by enzyme staining techniques (8) to be ATPase, is directly related to the increase in the complex of immunoprecipitates numbered 2. That this complex of precipitates represented the ATPase subunit(s) was confirmed by polyacrylamide gel electrophoresis of partially purified ATPase, extraction of the enzyme-active band with SDS, followed by CIE of the extract (329a; see Fig. 8,D). Combined SDS-polyacrylamide gel electrophoresis-CIE revealed that the two main ATPase subunits (α and β) were immunologically distinct (135a).

Further analysis of *M. lysodeikticus* ATPase by use of CIE with an intermediate gel showed that whereas native ATPase was capable of reacting with antisera raised to isolated membranes, to native ATPase, and to fully dissociated ATPase, the dissociated components reacted completely only with their homologous antiserum (Fig. 9). Reactions of inward feet were evident when intermediate gels contained high-titer antiserum to native ATPase, indicating low cross reactivity (see Fig. 9,B). Surprisingly, CIE of dissociated ATPase against antimembrane serum not only failed to yield a distinct precipitate but also prevented immunoprecipitation with the homologous antibody (Fig. 9,C). Possibly, ATPase subunit(s) form soluble immune complexes with antibodies in antimembrane serum that are then unable to react with homologous antibody for steric reasons. On the other hand, if ATPase subunits form soluble immune complexes with antibodies to native ATPase, blocking of the subsequent interaction with antibodies to dissociated ATPase does not occur (Fig. 9,B). Thus, it would appear that membrane-associated ATPase may possess determinants that are not expressed on the purified enzyme but that are expressed on the dissociated subunits of the partially purified

Triton X-100. Twelve micrograms of protein were analyzed by CIE in all instances in gels supplemented with 1% (v/v) Triton X-100. Gel regions marked a, b, c, and d contain agarose alone, antidissociated ATPase, antinative ATPase, and antimembrane serum, respectively. The concentrations of antiserum were adjusted to give similar titers against native ATPase. Precipitin complex 2 represents the ATPase subunits. It shows only reactions of inward feet when tested against antinative ATPase (compare A and B) and is totally absent from the precipitin pattern observed in C. Precipitate x represents the marker protein carbamylated transferrin, antibody to which was incorporated into gel b only. Other minor contaminating components are also evident. From Salton and Owen (329c).

enzyme. This phenomenon may reflect a conformational change in the enzyme after its release from the membrane or a loss of a minor subunit(s) on purification. Indeed, membrane-associated ATPase is largely in the form of a latent enzyme, which requires trypsin treatment for full expression of activity (330,331). A soluble form that possesses partial latency can be obtained by ionic shock treatment of washed membranes (273,328), and it may be significant that trypsin treatment of this preparation not only elicits full expression of ATPase activity but also increases the electrophoretic mobility of the enzyme in CIE experiments (Fig. 10). Purified ATPase, on the other hand, possesses little, if any, latency (272,330) and displays maximal electrophoretic mobility in CIE experiments. These phenomena may, again, reflect loss or alteration of a subunit(s) on trypsin treatment and purification. Indeed, recent evidence from SDS-polyacrylamide slab gel electrophoresis experiments has suggested that loss of latency may be related to degradation or loss of the ϵ-subunit (331a). It is uncertain at present, however, whether the different immunologic reactions observed for dissociated ATPase in Fig. 9 and for enzyme latency are truly related, because other explanations are certainly possible.

A note of caution should be introduced here with regard to the interpretation of reactions of full and partial identity in CIE. These remarks have special relevance in the study of enzyme substructure. Lines of partial identity can often occur for two identical antigens if precipitation of one species occurs before that of the other (153). We have recently shown that vertical displacement of wells in tandem CIE experiments by as little as 5 mm causes lines of partial identity between identical ATPase preparations. This figure is equivalent to a time displacement of only 20 min in the second-dimension electrophoresis (135a). In a similar manner, lines of partial identity may be observed for immunologically identical oligomeric forms of an enzyme if their electrophoretic mobilities differ significantly. In this respect, the lines of partial identity observed for oligomeric forms of β-amylase based on pore gradient polyacrylamide gel electrophoresis-CIE experiments should be interpreted with caution (200). It seems equally plausible that they result from differences in the relative rates of immunoprecipitation. The fact that β-mercaptoethanol conversion of the various preelectrophoresed oligomeric forms to monomers prior to second-dimension electrophoresis into antibody resulted in reactions of complete identity between components (200) is certainly compatible with this suggestion.

12. CLINICAL APPLICATIONS

Laurell (9,10,332) originally developed CIE to study serum proteins, particularly α_1-antitrypsin. Since then, CIE techniques have been used to study a wide variety of clinical problems, some of which have been covered earlier in the sections on complex formation and fragmentation. A detailed study of other areas is obviously outside the scope of this review, but problems of direct clinical importance concerning body-fluid enzymes and components of nonhuman origin, such as toxins and bacteria, are worthy of discussion.

Levels of human serum oxidase, ceruloplasmin, have been monitored extensively by CIE in the sera of normal adults (3,45,57,327,333–342,397), children (340,343,344), and newborns (339,345) and in the sera of patients suffering from a wide variety of

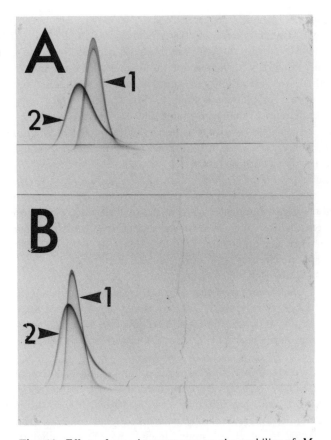

Fig. 10. Effect of trypsin treatment on the mobility of *M. lysodeikticus* ATPase. Partially purified ATPase (2.0 mg of protein/ml) was incubated for 3 hr at 23°C in the absence (A) and in the presence (B) of insolubilized trypsin (Enzyte-trypsin, 5 mg/ml). Supernatant fractions were assayed for ATPase activity and then mixed with carbamylated transfer-rin (46). Ten micrograms of each ATPase preparation were examined by CIE against antiserum that contained both anti-ATPase and antihuman transferrin. Precipitates 1 and 2 represent ATPase and carbamylated transferrin, respectively. ATPase preparations in A and B possessed 1.2 and 2.6 units of activity/ml, respectively, when assayed in the absence of soluble trypsin. The activity of both preparations was 2.6 units/ml when assayed in the presence of soluble trypsin. Note that the mobility of ATPase relative to that of carbamylated transferrin has increased after treatment with trypsin from 0.94 (A) to 0.98 (B). Relative mobilities as low as 0.86 have been detected for ATPase preparations that can be activated by trypsin by almost 300%. From Salton et al. (330a).

clinical conditions (3,213,327,336,337,339,340,343–368). Serum cholinesterase has been followed by CIE in only a few of these studies (335,340,343,350,367,368).

Johansson and Malmquist showed that the basic protein lysozyme could be quantitated conveniently by RIE with immunoplates arranged for cathodal migration (60). By use of this method, Johansson and colleagues were able to monitor increases in serum and urinary lysozyme for patients with monocytic leukemia (60), renal disorders (61), and for those undergoing renal transplants (369). Nørgaard-Pedersen and Mondrup also used RIE to quantitate the two carbonic anhydrases (B and C) of the human erythrocyte (48) and subsequently correlated levels of these enzymes with hyperthyroidism and with maturity of newborn infants (47).

RIE has also found use in the estimation of protease activity (321), in the quantitation of pepsin and pepsinogen in the gastric juices and urine of normal patients and those suffering from gastric diseases (63), in the quantitation of prostatic acid phosphatase from patients with suspected prostatic cancer (58), and in the detection of the brush-border marker enzyme alanine aminopeptidase in the urine of patients undergoing kidney transplant rejection (149) or acute tabular necrosis (59).

Sensitive RIE methods have also been devised for the detection of staphylococcal enterotoxin (77), *Clostridium perfringens* enterotoxin (40), and botulinum toxin (370) in contaminated foods. A reversed RIE system has been developed for the estimation of diphtheria and tetanus antibodies in human serum (89). CIE has been used to study α-amylases in sputum in an attempt to relate results with the clinical status of the patient (62).

CIE has greatly revolutionized the analysis of human antibody response to microbial infections. Most of the original work in this field was pioneered by Axelsen and coworkers studying *C. albicans* infections (14,25,140,232,371–373) and by Høiby and coworkers studying *P. aeruginosa* infections in patients suffering from cystic fibrosis (23,26,243,374–376). In a related study, Høiby and Wiik studied the occurrence of precipitins in the sera of patients with respiratory tract infections caused by *S. aureus*, *Hemophilus influenzae*, and *Diplococcus pneumoniae* (376). Similar experiments have also been performed recently in the study of patients with leprosy infections (377), actinomycosis (245), and chlamydial infections (162). A crossed radioimmunoelectrophoresis method has also been developed for the detection of antibodies to allergens (82,240,378).

13. CONCLUDING REMARKS

The methodology of quantitative immunoelectrophoresis has developed over the past years to a point where it can be used or adapted to investigate many aspects of enzymology. One of our primary aims has been to demonstrate this diversity. However, it seems fair to advise potential users of the technique to acquaint themselves thoroughly with the basic methodologic texts referred to in Sec. 1 (5,12–14), because a certain degree of dexterity is required, especially with micromethodology. Practical expertise can be gained with a little patience and perseverance.

It seems more than likely that quantitative immunoelectrophoresis will find increasing use in many areas of enzymology, such as in the direct estimation of enzyme content, in the molecular and subunit characterization of enzymes, in taxonomic and evolutionary aspects of comparative serology, and in comparative analysis of cell fractions. Indeed, in our studies (8,24), CIE has certainly proved to be more useful than SDS-polyacrylamide gel electrophoresis for probing the structural and functional relationships that exist within the bacterial membrane. The controversy that centers around the sidedness of isolated bacterial membrane vesicles (258) could well be resolved by CIE studies along parallel lines to those illustrated for *M. lysodeikticus* protoplasts (see Fig. 7 and Ref. 8).

Proteins, particularly of membrane origin, display marked dissimilarities with respect to solubilization by nonionic detergents. The denaturing effects of ionic detergents and their disruptive effects on immunoprecipitation have previously limited their use for immunochemical analysis. However, such methodologic problems have been tackled and solved in many different ways in several laboratories. The application of CIE techniques to the investigation of subunit structure of SDS-dissociated ATPase (Fig. 8) is one example of its potential. Although by no means technically easy, SDS-disc crossed immunoelectrophoresis may even permit immunogenicity of subunits to be related to that of the native enzyme. The possibility that quantitative immunoelectrophoretic methods could be used for definitive studies of the hydrophobic proteins and enzymes of the outer membrane of gram-negative bacteria is also very exciting. Localization of antigenic determinants, the active sites of enzymes, and binding sites of inhibitors, together with investigation of allosteric phenomena, are all potentially possible as well. Furthermore, the causes and nature of enzyme heterogeneity can be revealed by various combination techniques, such as crossed immunoelectrofocusing (69,70) and thin-layer gel filtration immunoelectrophoresis (199).

Enzymes have provided one basis for taxonomic and evolutionary studies in comparative biology, physiology, and biochemistry. For example, zymogram methods in combination with disc or slab polyacrylamide gel electrophoresis have been widely employed in bacterial taxonomy (398) and as indicators of evolutionary trends in speciation (399). Grouping of bacterial strains on the basis of such zymogram patterns alone, that is, where charge is the sole criterion of individuality, imposes certain limitations on the value of the devised classification scheme, because charge heterogeneity may be intrinsic, artifactual, or even dependent on cultural conditions (400). Crossed immunoelectrophoresis and related combination techniques could provide information on the immunologic relatedness of such heterogeneous components. Crossed-line immunoelectrophoresis or line immunoelectrophoresis (16,141,224) also provide alternative approaches for quantitation of immunologic cross reactivities, if the enzymes in question can be readily purified. Moreover, direct visualization of relatedness is more appealing than differential quantitative binding of labeled antibodies to a series of isofunctional proteins.

Quantitative immunoelectrophoresis, as with many techniques, not only provides answers but also poses several questions. The multienzyme complexes, which have been identified in rat liver plasma and microsomal membranes and which have been discussed at length in this chapter, provide one example. However, these phenomena have served to highlight the problems of interpreting enzyme staining patterns obtained for detergent-solubilized membrane antigens. Should these multienzyme complexes be shown to reflect an *in vivo* association of membrane antigens, CIE will prove to be a tool of immeasurable value in studying the complexity of functional mosaics that exist within the biologic membrane.

ACKNOWLEDGMENTS

The authors express their gratitude to Klaus Berzins and Jan Söderholm for kindly supplying Figs. 5 and 4, respectively, to Michael Huberman for permission to refer to and reproduce unpublished results concerning *M. lysodeikticus* ATPase, to Milton R. J. Salton for constructive criticism and helpful discussions, to Gillian Owen for proof reading the manuscript, and to Josephine Markiewicz for excellent secretarial service.

REFERENCES

1. C.-B. Laurell, *Scand. J. Clin. Lab. Invest.*, **29**(Suppl. 124), 21 (1972).
2. B. Weeke, *Scand. J. Immunol.*, **2**(Suppl. 1), 15 (1973).
3. B. Weeke, *Scand. J. Immunol.*, **2**(Suppl. 1), 37 (1973).
4. B. Weeke, *Scand. J. Immunol.*, **2**(Suppl. 1), 47 (1973).
5. R. Verbruggen, *Clin. Chem.*, **21**, 5 (1975).
6. P. Grabar and C. A. Williams, *Biochim. Biophys. Acta*, **10**, 193 (1953).
7. Y. Fukui, M. S. Nachbar, and M. R. J. Salton, *J. Bacteriol.*, **105**, 86 (1971).
8. P. Owen, and M. R. J. Salton, *Proc. Nat. Acad. Sci. USA*, **72**, 3711 (1975).
9. C.-B. Laurell, *Scand. J. Lab. Clin. Invest.*, **17**, 271 (1965).
10. C.-B. Laurell, *Anal. Biochem.*, **10**, 358 (1965).
11. C.-B. Laurell, Ed., *Electrophoretic and Electro-immuno-chemical Analysis of Proteins*, *Scand. J. Clin. Lab. Invest.* **29**(Suppl. 124), 1972.
12. N. H. Axelsen, J. Krøll, and B. Weeke, Eds., *A Manual of Quantitative Immunoelectrophoresis*, Universitetsforlaget, Osla, 1973.
13. N. H. Axelsen, Ed., *Quantitative Immunoelectrophoresis. New Development and Applications*, Universitetsforlaget, Osla, 1975.
14. N. H. Axelsen, in C.-G. Hedén and T. Illéni, Eds., *Automation in Microbiology and Immunology*, John Wiley and Sons, New York, 1975, p. 356.
15. O. J. Bjerrum and T. C. Bøg-Hansen, in A. H. Maddy, Ed., *Biochemical Analysis of Membranes*, Chapman and Hall, London, 1976.
16. E. Bock and N. H. Axelsen, *J. Immunol. Methods*, **2**, 75 (1973).
17. N. H. Axelsen, *Infect. Immunity*, **7**, 949 (1973).
18. O. J. Bjerrum, *Int. J. Biochem.*, **6**, 513 (1975).
19. N. Harboe and A. Ingild, *Scand. J. Immunol.*, **2**(Suppl. 1), 161 (1973).
20. P.-O. Ganrot, *Scand. J. Clin. Lab. Invest.*, **29**(Suppl. 124), 67 (1972).
21. J. H. Humphrey and R. G. White, *Immunology for Students of Medicine*, 3rd edit., Blackwell Scientific Publishing Co., Oxford, 1970.
22. N. Høiby, *Scand. J. Immunol.*, **4**(Suppl. 2), 187 (1975).
23. N. Høiby, *Scand. J. Immunol.*, **4**(Suppl. 2), 197 (1975).
24. C. J. Smyth, A. E. Friedman-Kien, and M. R. J. Salton, *Infect. Immunity*, **13**, 1273 (1976).
25. N. H. Axelsen and P. J. Svendsen, *Protides Biol. Fluids*, **19**, 561 (1972).
26. N. Hoiby and N. H. Axelsen, *Acta Pathol. Microbiol. Scand. Sect. B*, **81**, 298 (1973).
27. O. Closs, M. Harboe, and A. M. Wassum, *Scand. J. Immunol.*, **4**(Suppl. 2), 173 (1975).
28. R. E. Syverson, H. R. Buckley, and C. C. Campbell, *Infect. Immunity*, **12**, 1184 (1975).
29. C.-H. Brogren, T. C. Bøg-Hansen, P. Just Svendsen, and O. J. Bjerrum, *Scand. J. Immunol.*, **4** (Suppl. 2), 59 (1975).
30. R. Verbruggen, *Biochem. J.*, **151**, 149 (1975).
31. K. Ohlsson, *Biochim. Biophys. Acta*, **236**, 84 (1971).
32. K. Ohlsson, *Clin. Chim. Acta*, **32**, 215 (1971).

33. K. Ohlsson, *Scand. J. Clin. Lab. Invest.,* **28,** 219 (1971).
34. O. J. Bjerrum, J. Ramlau, I. Clemmesen, A. Ingild, and T. C. Bøg-Hansen, *Scand. J. Immunol.,* **4**(Suppl. 2), 81 (1975).
35. O. J. Bjerrum, P. Lundhal, C.-H. Brogren, and S. Hjertén, *Biochim. Biophys. Acta,* **394,** 173 (1975).
36. O. J. Bjerrum, *Experientia,* **30,** 831 (1974).
37. T. C. Bøg-Hansen and O. J. Bjerrum, *Protides Biol. Fluids,* **21,** 39 (1974).
38. M. Hughes, R. O. Thomson, P. Knight, and J. Stephen, *J. Appl. Bacteriol.,* **37,** 603 (1974).
39. K. B. Guiseley and D. W. Renn, *Agarose: Purification, Properties and Biomedical Applications,* Marine Colloids, Inc., Rockland, Maine, 1975.
40. C. L. Duncan and E. B. Somers, *Appl. Microbiol.,* **24,** 801 (1972).
41. A. O. Grubb, *J. Immunol.,* **112,** 1420 (1974).
42. R. J. Wieme, *Agar Gel Electrophoresis,* Elsevier, Amsterdam, 1965.
43. V. Milisauskas and N. R. Rose, *Exp. Cell Res.,* **81,** 279 (1973).
44. W. Von Stephan and U. Frahm, *Z. Klin. Chem. Klin. Biochem.,* **9,** 224 (1971).
45. H. G. M. Clarke and T. Freeman, *Clin. Sci.,* **35,** 403 (1968).
46. B. Weeke, *Scand. J. Clin. Lab. Invest.,* 25, **161** (1970).
47. B. Nørgaard-Pedersen, *Scand. J. Immunol.,* **2**(Suppl. 1), 125 (1973).
48. B. Nørgaard-Pedersen and M. Mondrup, *Scand. J. Clin. Lab. Invest.,* **27,** 169 (1971).
49. L. Slater, *Ann. Clin. Biochem.,* **12,** 19 (1975).
50. P. Lundhal, *Biochim. Biophys. Acta,* **379,** 304 (1975).
51. S. Bhakdi, H. Knüfermann, H. Fischer, and D. F. H. Wallach, *Biochim. Biophys. Acta,* **373,** 295 (1974).
52. F. Wengenmayer, K.-H. Ueberschär, and G. Kurz, *FEBS Lett.,* **40,** 224 (1974).
53. M. Hiebert, J. Gauldie, and B. L. Hillcoat, *Anal. Biochem.,* **46,** 433 (1972).
54. J. M. B. Versey, J. R. Hobbs, and P. J. L. Holt, *Ann. Rheumatic Diseases,* **32,** 557 (1973).
55. C. E. Thirkhill and G. E. Kenny, *J. Immunol.,* **114,** 1107 (1975).
56. D. Merrill, T. F. Hartley, and H. N. Claman, *J. Lab. Clin. Med.,* **69,** 151 (1967).
57. C. W. Gill, C. L. Fischer, and C. L. Holleman, *Clin. Chem.,* **17,** 501 (1971).
58. V. Milisauskas and N. R. Rose, *Clin. Chem.,* **18,** 1529 (1972).
59. J. E. Scherberich, F. W. Falkenberg, A. W. Mondorf, H. Müller, and G. Pfleiderer, *Clin. Chim. Acta,* **55,** 179 (1974).
60. B. G. Johansson and J. Malmquist, *Scand. J. Clin. Lab. Invest.,* **27,** 255 (1971).
61. B. G. Johansson and U. Rovnskov, *Scand. J. Urol. Nephrol.,* **6,** 249 (1972).
62. H. C. Ryley, *Biochim. Biophys. Acta,* **271,** 300 (1972).
63. H. Hirsch-Marie, *Clin. Chim. Acta,* **24,** 411 (1969).
64. J. Daussant, J. Roussaux, and P. Manigault, *FEBS Lett.,* **14,** 245 (1971).
65. J. Heigaard and T. C. Bøg-Hansen, *J. Inst. Brewing,* **80,** 436 (1974).
66. T. C. Bøg-Hansen and J. Daussant, *Anal. Biochem.,* **61,** 522 (1974).
67. A. R. Bradwell and D. Burnett, *Clin. Chim. Acta,* **58,** 283 (1975).
68. J.-E. Nilêhn, *Thromb. Diath. Haemorrhag.,* **18,** 89 (1967).
69. J. Söderholm and C. J. Smyth, in P. G. Righetti, Ed., *Progress in Isoelectric Focusing and Isotachophresis,* North-Holland Publishing Company, Amsterdam, 1975, p. 99.
70. J. Söderholm, C. J. Smyth, and T. Wadström, *Scand. J. Immunol.,* **4**(Suppl. 2), 107 (1975).
71. K. Berzins, F. Blomberg, M. Kjellgren, C. J. Smyth, and T. Wadström, *FEBS Lett.,* **61,** 77 (1976).
72. S. Bach, S. Ruddy, J. A. McLaren, and K. F. Austen, *Immunology,* **21,** 869 (1971).
73. G. Kostner and A. Holasek, *Anal. Biochem.,* **46,** 680 (1972).
74. K. Ganrot, *Biochim. Biophys. Acta,* **295,** 245 (1973).
75. N. St. G. Hyslop and D. G. Cochrane, *J. Immunol. Methods,* **6,** 99 (1974).
76. D. A. Darcy, *Clin. Chim. Acta,* **38,** 329 (1972).
77. E. Gasper, R. C. Heimsch, and A. W. Anderson, *Appl. Microbiol.,* **25,** 421 (1973).
78. R. P. Propp, B. Jabbari, K. Barron, and C. Wiggins, *J. Lab. Clin. Med.,* **82,** 154 (1973).
79. C. A. Saravis and L. Bonacker, *Nature (London),* **228,** 61 (1970).
80. G. Grieninger and S. Granick, *Proc. Nat. Acad. Sci. USA,* **72,** 5007 (1975).
81. T. Freeman and J. D. Pearson, *Clin. Chim. Acta,* **26,** 365 (1969).

82. B. Weeke and H. Løwenstein, *Scand. J. Immunol.,* **2**(Suppl. 1), 149 (1973).
83. C.-O. Kindmark and J. I. Thorell, *Scand. J. Clin. Lab. Invest.,* **29** (Suppl. 124), 49 (1972).
84. C.-H. Brogren and T. C. Bøg-Hansen, *Scand. J. Immunol.,* **4**(Suppl. 2), 37 (1975).
85. O. J. Bjerrum, A. Ingild, H. Løwenstein, and B. Weeke, *Scand. J. Immunol.,* **2**(Suppl. 1), 145 (1973).
86. A. J. Crowle, G. J. Revis, and K. Jarrett, *Immunol. Commun.,* **1,** 325 (1972).
87. N. H. Axelsen and P. J. Svendsen, *Scand. J. Immunol.,* **2**(Suppl. 1), 155 (1973).
88. R. J. Collier, *Bacteriol. Rev.,* **39,** 54 (1975).
89. H. M. Höyeraal, B. Vandvik, and O. J. Mellbye, *J. Immunol. Methods,* **6,** 385 (1975).
90. S. Fleischer and L. Packer, Eds., *Biomembranes, Methods Enzymol.* **31,**(A), 1974.
91. A. Helenius and K. Simons, *Biochem. Biophys. Acta,* **415,** 29 (1975).
92. R. Schmidt-Ullrich, D. F. H. Wallach, and J. Hendricks, *Biochim. Biophys. Acta,* **382,** 295 (1975).
93. O. J. Bjerrum and P. Lundahl, *Scand. J. Immunol.,* **2**(Suppl. 1), 139 (1973).
94. F. Blomberg, Enzyme-Active Membrane Antigens in Rat Liver, Ph.D. Thesis, University of Stockholm, Sweden, 1974.
95. O. J. Bjerrum and P. Lundahl, *Biochim. Biophys. Acta,* **342,** 69 (1974).
96. K. Berzins, F. Blomberg, and P. Perlmann, *Eur. J. Biochem.,* **51,** 181 (1975).
97. C. S. Nielsen and O. J. Bjerrum, *Scand. J. Immunol.,* **4**(Suppl. 2), 73 (1975).
98. T. V. Waehneldt, *Biosystems,* **6,** 176 (1975).
99. O. J. Bjerrum and T. C. Bøg-Hansen, *Scand. J. Immunol.,* **4**(Suppl. 2), 89 (1975).
100. H. Loft, *Scand. J. Immunol.,* **4**(Suppl. 2), 115 (1975).
101. E. Bock, E. T. Mellerup, and O. J. Rafaelsen, *J. Neurochem.,* **18,** 2435 (1971).
102. L. Nielsen, H. Løwenstein, and B. Weeke, *Acta Allergol.,* **29,** 385 (1974).
103. N. Høiby, *Acta Pathol. Microbiol. Scand. Sect. B,* **83,** 433 (1975).
104. Z. A. Tökes and S. M. Chambers, *Biochim. Biophys. Acta,* **389,** 325 (1975).
105. M. Heller, P. Edelstein, and M. Mayer, *Biochim. Biophys. Acta,* **413,** 472 (1975).
106. C. A. Nelson, *J. Biol. Chem.,* **246,** 3895 (1971).
107. J. R. Pringle, *Biochem. Biophys. Res. Commun.,* **39,** 46 (1970).
108. B. Weeke, *Scand. J. Lab. Clin. Invest.,* **25,** 269 (1970).
109. N. Høiby, *Acta Pathol. Microbiol. Scand. Sect. B,* **83,** 321 (1975).
110. B. Cinader, *Annu. Rev. Microbiol.,* **11,** 371 (1957).
111. B. Cinader, *Ann. N.Y. Acad. Sci.,* **103,** 495 (1963).
112. B. Cinader, Ed., in *Antibodies to Biologically Active Molecules,* Pergamon Press, Oxford, 1967, p. 85.
113. R. Arnon, *Curr. Topics Microbiol. Immunol.,* **54,** 47 (1971).
114. R. Arnon, in M. Sela, Ed., *The Antigens,* Vol. 1, Academic Press, New York, 1973, p. 88.
115. P. H. Maurer, in C. A. Williams and M. W. Chase, Eds., *Methods in Immunology and Immunochemistry,* Vol. 3, Academic Press, New York, 1971, p. 1.
116. J. R. Cann, *Immunochemistry,* **12,** 473 (1975).
117. J. R. Cann, *Biophys. Chem.,* **3,** 206 (1975).
118. J. D. Oppenheim and M. S. Nachbar, this volume (Chap. 4).
119. J. Uriel, *Ann. N.Y. Acad. Sci.,* **103,** 956 (1963).
120. J. Uriel, in B. Cinader, Ed., *Antibodies to Biologically Active Molecules,* Pergamon Press, Oxford, 1967, p. 181.
121. J. Uriel, in C. A. Williams and M. W. Chase, Eds., *Methods in Immunology and Immunochemistry,* Vol. 3, Academic Press, New York, 1971, p. 294.
122. R. H. Maurer, *Disc Electrophoresis and Related Techniques of Polyacrylamide Gel Electrophoresis,* 2nd edit., De Gruyter, Berlin, 1971.
123. B. Radola, in J. P. Arbuthnott and J. A. Beeley, Eds., *Isoelectric Focusing,* Butterworths, London, 1975, p. 182.
124. T. Wadström and C. J. Smyth, in J. P. Arbuthnott and J. A. Beeley, Eds., *Isoelectric Focusing,* Butterworths, London, 1975, p. 152.
125. E. Bock and J. Dissing, *Scand. J. Immunol.,* **4**(Suppl. 2), 31 (1975).
126. F. Blomberg and K. Berzins, *Eur. J. Biochem.,* **56,** 319 (1975).
127. H. Suginaka, P. M. Blumberg, and J. L. Strominger, *J. Biol. Chem.,* **247,** 5279 (1972).
128. P. M. Blumberg and J. L. Strominger, *Bacteriol. Rev.,* **38,** 291 (1974).

129. J. J. Pollock, M. Nguyen-Distèche, J.-M. Ghuysen, J. Coyette, R. Linder, M. R. J. Salton, K. S. Kim, H. R. Perkins, and P. Reynolds, *Eur. J. Biochem.*, **41,** 439 (1974).

130. B. L. Vallee, *Advan. Protein Chem.*, **10,** 318 (1955).

131. P. O. Ganrot and J.-E. Niléhn, *Scand. J. Clin. Lab. Invest.*, **21,** 238 (1968).

132. J. Stenflo, *Acta Chem. Scand.*, **24,** 3762 (1970).

133. F. Blomberg and M. Raftell, *Eur. J. Biochem.*, **49,** 21 (1974).

134. M. Raftell and C. Powell, *Immunochemistry*, **7,** 619 (1970).

135. T. C. Bøg-Hansen, H. Brogren, and M. Rostgaard, *Int. Res. Commun. Syst.*, **73-9,** 3122 (1973).

135a. P. Owen and M. R. J. Salton, Unpublished results (1975).

136. M. Raftell, F. Blomberg, and P. Perlmann, *Cancer Res.*, **34,** 2300 (1974).

137. F. Blomberg and P. Perlmann, *Exp. Cell Res.*, **66,** 104 (1971).

138. IUPAC-IUB Commission, *Arch. Biochem. Biophys.* **147,** 1 (1971).

139. B. F. Vestergaard and P. C. Grauballe, *Scand. J. Immunol.*, **4**(Suppl. 2), 207 (1975).

140. N. H. Axelsen and E. Bock, *J. Immunol. Methods,* **1,** 109 (1972).

141. A. Grubb, *Scand. J. Clin. Lab. Invest.*, **29**(Suppl. 124), 59 (1972).

142. N. H. Axelsen, E. Bock, and J. Kroll, *Scand. J. Immunol.*, **2**(Suppl. 1), 91 (1973).

143. E. Bock and N. H. Axelsen, *Scand. J. Immunol.*, **2**(Suppl. 1), 95 (1973).

144. P. J. Svendsen, *Scand. J. Immunol.*, **2**(Suppl. 1), 69 (1973).

145. K. Berzins and F. Blomberg, *FEBS Lett.*, **54,** 139 (1975).

146. J. Krøll, *Scand. J. Immunol.*, **2**(Suppl. 1), 61 (1973).

147. J. Krøll, *Scand. J. Immunol.*, **2**(Suppl. 1), 83 (1973).

148. J. Krøll, *Scand. J. Immunol.*, **2**(Suppl. 1), 57 (1973).

149. A. W. Mondorf, C. B. Carpenter, J. E. Scherberich, and J. P. Merrill, *Protides Biol. Fluids*, **21,** 493 (1974).

150. D. Herbold, R. Zwilling, and G. Pfleiderer, *Hoppe-Seyler's Z. Physiol. Chem.*, **352,** 583 (1971).

151. O. Notén, H. Sjöström, and L. Josefsson, *Biochim. Biophys. Acta,* **327,** 446 (1973).

152. H. Sjöström, O. Notén, and L. Josefsson, *Biochim. Biophys. Acta,* **327,** 457 (1973).

153. J. Krøll, *Scand. J. Immunol.*, **2**(Suppl. 1), 79 (1973).

154. H. Løwenstein, O. J. Bjerrum, E. Weeke, and B. Weeke, *Scand. J. Immunol.*, **4**(Suppl. 2), 155 (1975).

155. O. J. Bjerrum, S. Bhakdi, T. C. Bøg-Hansen, H. Knüfermann, and D. F. H. Wallach, *Biochim. Biophys. Acta,* **406,** 489 (1975).

156. W. Giebel and H. Saechtling, *Hoppe-Seyler's Z. Physiol. Chem.*, **354,** 673 (1973).

157. N. H. Axelsen, *Scand. J. Immunol.*, **2**(Suppl. 1), 71 (1973).

158. M. M. Andersen and J. Krøll, *Scand. J. Immunol.*, **4**(Suppl. 2), 163 (1975).

159. J. Krøll and M. M. Andersen, *J. Immunol. Methods,* **9,** 141 (1975).

160. W. Groc and M. Jendrey, *Clin. Chim. Acta,* **52,** 59 (1974).

161. M. P. Ollier and L. Hartman, *Biomedicine,* **21,** 444 (1974).

162. H. D. Caldwell, C.-C. Kuo, and G. E. Kenny, *J. Immunol.*, **115,** 969 (1975).

163. B. F. Vestergaard, *Scand. J. Immunol.*, **4**(Suppl. 2), 203 (1975).

164. C. Koch and H. E. Nielsen, *Scand. J. Immunol.*, **4**(Suppl. 2), 121 (1975).

165. T. C. Bøg-Hansen and M. Nord, *J. Biol. Educ.*, **8,** 167 (1974).

166. T. C. Bøg-Hansen, C.-H. Brogren, and I. McMurrough, *J. Inst. Brewing,* **80,** 443 (1974).

167. P. Owen and M. R. J. Salton, *Anal. Biochem.*, **73,** 20 (1976).

168. T. C. Bøg-Hansen, *Anal. Biochem.*, **56,** 480 (1973).

169. T. C. Bøg-Hansen, O. J. Bjerrum, and J. Ramlau, *Scand. J. Immunol.*, **4**(Suppl. 2), 141 (1975).

170. M. Harboe, E. Saltvedt, O. Closs, and S. Olsnes, *Scand. J. Immunol.*, **4**(Suppl. 2), 125 (1975).

170a. P. Owen and J. D. Oppenheim, Unpublished results (1975).

171. B. Weeke, E. Weeke, and H. Lowenstein, *Scand. J. Immunol.*, **4**(Suppl. 2), 149 (1975).

172. B. F. Vestergaard and T. C. Bøg-Hansen, *Scand. J. Immunol.*, **4**(Suppl. 2), 211 (1975).

173. T. C. Bøg-Hansen and C.-H. Brogren, *Scand. J. Immunol.*, **4**(Suppl. 2), 135 (1975).

174. N. Catsimpoolas, *Ann. N.Y. Acad. Sci.*, **209,** 144 (1973).

175. O. Vesterberg, *Sci. Tools,* **20,** 22 (1973).

176. K. Ekwall, J. Söderholm, and T. Wadström, *J. Immunol. Methods,* in press (1976).
177. N. Ressler, *Clin. Chim. Acta,* **5,** 795 (1960).
178. M. K. Fagerhol and C.-B. Laurell, *Clin. Chim. Acta,* **16,** 199 (1967).
179. A. Grubb, *Anal. Biochem.* **55,** 582 (1973).
180. A. Grubb, *Protides Biol. Fluids,* **21,** 649 (1974).
181. B. G. Johansson and S. Hjertén, *Anal. Biochem.,* **59,** 200 (1974).
182. H. C. Weise and D. Graesslin, in P. G. Righetti, Ed., *Progress in Isoelectric Focusing and Isotachophoresis,* North Holland Publishing Company, Amsterdam, 1975, p. 93.
183. B. G. Johansson and J. Stenflo, *Anal. Biochem.,* **40,** 232 (1971).
184. C.-B. Laurell and U. Persson, *Biochim. Biophys. Acta,* **310,** 500 (1973).
185. W. Dames, H. R. Maurer, and V. Neuhoff, *Hoppe-Seyler's Z. Physiol. Chem.,* **353,** 554 (1972).
186. A. M. Johnson, K. Schmid, and C. A. Alper, *J. Clin. Invest.,* **48,** 2293 (1969).
187. J. Uriel, *Bull. Soc. Chim. Biol.,* **48,** 969 (1966).
188. G. Skude and J. O. Jeppsson, *Scand. J. Clin. Lab. Invest.,* **29**(Suppl. 124), 55 (1972).
189. P. Lundahl and L. Liljas, *Anal. Biochem.,* **65,** 50 (1975).
190. R. L. Juliano, *Biochim. Biophys. Acta,* **300,** 341 (1973).
191. A. L. Shapiro, E. Viñuela, and J. V. Maizel, Jr., *Biochem. Biophys. Res. Commun.,* **28,** 815 (1967).
192. K. Weber and M. Osborn, *J. Biol. Chem.,* **244,** 4406 (1969).
193. A. K. Dunker and R. R. Rueckert, *J. Biol. Chem.,* **244,** 5074 (1969).
194. C. A. Converse and D. S. Papermaster, *Science,* **189,** 469 (1975).
195. S. Clarke, *J. Biol. Chem.,* **250,** 5459 (1975).
196. G. P. Tuszynski and L. Warren, *Anal. Biochem.,* **67,** 55 (1975).
197. K. Weber and D. J. Kuter, *J. Biol. Chem.,* **246,** 4504 (1971).
198. J. Lenard, *Biochem. Biophys. Res. Commun.,* **45,** 662 (1971).
199. C. S. Nielsen, *Scand. J. Immunol.,* **4**(Suppl. 2), 101 (1975).
200. J. Daussant and A. Skakoun, *J. Immunol. Methods,* **7,** 39 (1975).
201. M. Jirka and P. Blanický, *Biochim. Biophys. Acta,* **295,** 1 (1973).
202. L. Belin, *Int. Arch. Allergy Appl. Immunol.,* **42,** 329 (1972).
203. J. W. Gurd, W. H. Evans, and H. R. Perkins, *Biochem. J.,* **135,** 827 (1973).
204. C. J. Smyth, *J. Gen. Microbiol.,* **87,** 219 (1975).
205. G. Skude, *Hereditas,* **65,** 277 (1970).
206. W. Giebel, R. Zwilling, and G. Pfleiderer, *Comp. Biochem. Physiol.,* **38B,** 197 (1971).
207. K. Ohlsson, *Biochim. Biophys. Acta,* **251,** 450 (1971).
208. D. Bout, J. Fruit, and A. Capron, *C. R. Acad. Sci. Paris Ser. D,* **276,** 2341 (1973).
209. R. Möllby and T. Wadström, *Biochim. Biophys. Acta,* **321,** 569 (1973).
210. K. Ohlsson and I. Olsson, *Eur. J. Biochem.,* **36,** 473 (1973).
211. S. Neumann, F. Falkenberg, and G. Pfleiderer, *Biochim. Biophys. Acta,* **334,** 328 (1974).
212. K. Ohlsson and I. Olsson, *Eur. J. Biochem.,* **42,** 519 (1974).
213. W. H. Evans, P. O. Hood, and J. W. Gurd, *Biochem., J.,* **135,** 819 (1975).
214. J. Stenflo, *Scand. J. Clin. Lab. Invest.,* **29**(Suppl. 126), 4.4 (Abst.) (1972).
215. T. C. Bøg-Hansen, O. J. Bjerrum, and P. J. Svendsen, *Sci. Tools,* **21,** 33 (1974).
216. H. Løwenstein, L. Nielsen, and B. Weeke, *Acta Allergol.,* **29,** 418 (1974).
217. O. Söderlind, R. Möllby, and T. Wadström, *Zentr. Bakteriol. Parasitenk. Abt. I Orig. A,* **229,** 190 (1974).
218. H. Løwenstein, L. Neilsen, and B. Weeke, *Int. Arch. Allergy Appl. Immunol.,* **49,** 95 (1975).
219. C.-H. Brogren and P. J. Svendsen, *Protides Biol. Fluids,* **22,** 685 (1975).
220. P. Owen and J. H. Freer, *Biochem. J.,* **120,** 237 (1970).
221. A. A. Yunis, G. K. Arimura, H. G. Haines, R. J. Ratzan, and M. A. Gross, *Cancer Res.,* **35,** 337 (1975).
222. S. Arvidson, *Acta Pathol. Microbiol. Scand. Sect. B,* **81,** 545 (1973).
223. R. Linke, E. D. Wachsmuth, and G. Pfleiderer, *Immunochemistry,* **7,** 99 (1970).
224. N. H. Axelsen and E. Bock, *J. Immunol. Methods,* **2,** 393 (1973).
225. S. Arvidson, *Acta Pathol. Microbiol. Scand. Sect. B,* **81,** 552 (1973).
226. F. Blomberg and M. Raftell, *Biochim. Biophys. Acta,* **291,** 431 (1973).

227. U. Lundkvist and P. Perlman, *Immunology,* **13,** 179 (1967).

228. M. Raftell and C. Powell, *Biochim. Biophys. Acta,* **216,** 428 (1970).

229. M. Raftell and F. Blomberg, *Eur. J. Biochem.,* **49,** 31 (1974).

230. F. Blomberg and P. Perlman, *Biochim. Biophys. Acta,* **233,** 53 (1971).

231. M. Raftell and F. Blomberg, *Biochim. Biophys. Acta,* **291,** 442 (1973).

232. N. H. Axelsen, H. R. Buckley, E. Drouhet, E. Budtz-Jørgensen, T. Hattel, and P. L. Anderson, *Scand. J. Immunol.,* **4**(Suppl. 2), 217 (1975).

233. P. Larsson, L. A. Hanson, and B. Kaijser, *Acta Pathol. Microbiol. Scand. Sect. B,* **81,** 641 (1973).

234. V. R. Linke, R. Zwilling, D. Herbold, and G. Pfleiderer, *Hoppe-Seyler's Z. Physiol. Chem.,* **350,** 877 (1969).

235. D. B. Roberts, G. L. Wright, L. F. Affronti, and M. Reich, *Infect. Immunity,* **6,** 564 (1972).

236. G. L. Wright and D. B. Roberts, *Amer. Rev. Resp. Diseases,* **109,** 306 (1974).

237. E. Bock, *J. Neurochem.,* **19,** 1731 (1972).

238. O. S. Jorgensen and E. Bock, *J. Neurochem.,* **23,** 879 (1974).

239. E. Bock, *Scand. J. Immunol.* **2**(Suppl. 1), 119 (1973).

240. B. Weeke and H. Løwenstein, *Int. Arch. Allergy Appl. Immunol.,* **49,** 74 (1975).

241. G. Pfleiderer and R. Zwilling, *Naturwissenschaften,* **59,** 396 (1972).

242. T. Wadström, M. Thelestam, and R. Möllby, *Ann. N.Y. Acad. Sci.,* **236,** 343 (1974).

243. N. Høiby, *Acta Pathol. Microbiol. Scand. Sect. B,* **83,** 328 (1975).

244. K. Holmberg, C.-E. Nord, and T. Wadström, *Infect. Immunity,* **12,** 387 (1975).

245. K. Holmberg, C.-E. Nord, and T. Wadström, *Infect. Immunity,* **12,** 398 (1975).

246. H. D. Caldwell, C.-C. Kuo, and G. E. Kenny, *J. Immunol.,* **115,** 963 (1975).

247. G. H. Sweet, D. E. Wilson, and J. D. Gerber, *J. Immunol.,* **111,** 554 (1973).

248. S. Jeansson and B. F. Vestergaard, *Acta Pathol. Microbiol. Scand. Sect. B,* **83,** 343 (1975).

249. B. F. Vestergaard, *Acta Pathol. Microbiol. Scand. Sect. B,* **81,** 808 (1973).

250. E. G. V. Evans, M. D. Richardson, F. C. Odds, and K. T. Holland, *Brit. Med. J.,* **4,** 86 (1973).

251. M. R. J. Salton, *Trans. N.Y. Acad. Sci. Ser. II,* **29,** 764 (1967).

252. M. R. J. Salton and M. S. Nachbar, in N. K. Boardman, A. W. Linnane, and R. M. Smillie, Eds., *Autonomy and Biogenesis of Mitochondria and Chloroplasts,* North Holland, Amsterdam, 1970, p. 42.

253. S. Shikbo and A. L. Tappel, *Biochim. Biophys. Acta,* **73,** 76 (1963).

254. F. M. Baccino, G. A. Rita, and M. F. Zuretti, *Biochem. J.,* **122,** 363 (1971).

255. E. Bock, O. S. Jorgensen, and S. J. Morris, *J. Neurochem.,* **22,** 1013 (1974).

256. S. Bhakdi, O. J. Bjerrum, U. Rother, H. Knüfermann, and D. F. H. Wallach, *Biochim. Biophys. Acta,* **406,** 21 (1975).

257. A. Solyom and E. G. Trams. *Enzyme,* **13,** 329 (1972).

258. M. R. J. Salton and P. Owen, *Annu. Rev. Microbiol.,* in press (1976).

259. B. K. Ghosh, *Sub-Cell. Biochem.,* **3,** 311 (1974).

260. M. R. J. Salton, *CRC Crit. Rev. Microbiol.,* **1,** 161 (1971).

261. V. M. Reusch and M. M. Burger, *Biochim. Biophys. Acta,* **300,** 79 (1974).

262. P. Owen and J. H. Freer, *Biochem. J.,* **129,** 907 (1972).

263. P. Owen and M. R. J. Salton, *Biochim. Biophys. Acta,* **406,** 214 (1975).

264. E. Bock and O. S. Jørgensen, *FEBS Lett.,* **52,** 37 (1975).

265. W. P. Hurlbert and B. Ceccarelli, *Advan. Cytopharmacol.,* **2,** 141 (1974).

266. S. J. Singer, *Annu. Rev. Biochem.,* **43,** 805 (1974).

267. J. D. Oppenheim and M. R. J. Salton, *Biochim. Biophys. Acta,* **298,** 297 (1973).

268. T. L. Steck and J. A. Kant, *Methods Enzymol.,* **31,** 172 (1974).

269. M. Futai, *J. Membrane Biol.,* **15,** 15 (1974).

270. M. Raftell and P. Perlman, *Exp. Cell Res.* **57,** 119 (1969).

271. J. W. Depierre and G. Dallner, *Biochim. Biophys. Acta,* **415,** 411 (1975).

272. M. R. J. Salton and M. T. Schor, *Biochim. Biophys. Acta,* **345,** 74 (1974).

273. A. Ginsburg and E. R. Stadtman, *Annu. Rev. Biochem.* **39,** 429 (1970).

274. L. J. Reed and D. J. Cox, *Annu. Rev. Biochem.,* **35,** 57 (1965).

275. P. F. Lue and J. G. Kaplan, *Biochem. Biophys. Res. Commun.,* **34,** 426 (1969).

276. P. F. Lue and J. G. Kaplan, *Biochim. Biophys. Acta,* **220,** 365 (1970).
277. P. F. Lue and J. G. Kaplan, *Can. J. Biochem.,* **48,** 155 (1970).
278. P. F. Lue and J. G. Kaplan, *Can. J. Biochem.,* **49,** 403 (1971).
279. B. Simon and L. Thomas, *Biochim. Biophys. Acta,* **288,** 434 (1972).
280. D. M. Aitkin, P. F. Lue, and J. G. Kaplan, *Can. J. Biochem.,* **53,** 721 (1975).
281. M. Mori, H. Ishida, and M. Tatibana, *Biochemistry,* **14,** 2622 (1975).
282. J. S. Wolpert and M. L. Ernst-Fonberg, *Biochemistry,* **14,** 1095 (1975).
283. J. S. Wolpert and M. L. Ernst-Fonberg, *Biochemistry,* **14,** 1103 (1975).
284. Y. Kagawa, *Biochim. Biophys. Acta,* **265,** 297 (1972).
285. H. Baltscheffsky and M. Baltscheffsky, *Annu. Rev. Biochem.,* **43,** 871 (1974).
286. D. E. Green, *Biochim. Biophys. Acta,* **346,** 27 (1974).
287. H. J. Harman, J. D. Hall, and F. L. Crane, *Biochim. Biophys. Acta,* **344,** 119 (1974).
288. N. S. Gel'man, M. A. Lukoyanova, and D. N. Ostrovskii, *Bacterial Membranes and the Respiratory Chain, Biomembranes,* Vol. 6, Plenum Press, New York, 1975.
289. P. O. Ganrot, *Scand. J. Clin. Lab. Invest.* **29,**(Suppl. 124), 39 (1972).
290. A. O. Grubb, *Scand. J. Immunol.,* **4**(Suppl. 2), 53 (1975).
291. A. H. Christiansen and J. Krøll, *Scand. J. Immunol.,* **2**(Suppl. 1), 133 (1973).
292. K. Ohlsson and H. Tegner, *Eur. J. Clin. Invest.,* **5,**221 (1975).
293. K. Ohlsson, *Scand. J. Clin. Lab. Invest.,* **28,** 5 (1971).
294. K. Ohlsson, *Scand. J. Clin. Lab. Invest.,* **28,** 225 (1971).
295. K. Ohlsson, *Scand. J. Clin. Lab. Invest.,* **28,** 251 (1971).
296. K. Ohlsson and I. Olsson, *Scand. J. Clin. Lab. Invest.,* **34,** 349 (1974).
297. K. Ohlsson, P.-O. Ganrot, and C.-B. Laurell, *Acta Chir. Scand.,* **137,** 113 (1971).
298. K. Ohlsson, *Scand. J. Clin. Lab. Invest.,* **28,** 13 (1971).
299. K. Ohlsson, *Scand. J. Gastroenterol.,* **6,** 645 (1971).
300. K. Ohlsson, *Clin. Chim. Acta,* **32,** 399 (1971).
301. K. Ganrot, *Biochim. Biophys. Acta,* **322,** 62 (1973).
302. J.-E. Niléhn and P. O. Ganrot, *Scand. J. Clin. Lab. Invest.,* **20,** 113 (1967).
303. S. Müllertz, *Scand. J. Clin. Lab. Invest.,* **30,** 369 (1972).
304. S. Müllertz, *Biochem. J.,* **143,** 273 (1974).
305. G. Sas, D. S. Pepper, and J. D. Cash, *Thromb. Res.,* **6,** 87 (1975).
306. H. Neurath, *Advan. Protein Chem.,* **12,** 319 (1957).
307. M. Yamasaki, J. R. Brown, D. J. Cox, R. N. Greenshield, R. N. Wade, and H. Neurath, *Biochemistry,* **2,** 859 (1963).
308. J. T. Barrett, *Int. Arch. Allergy Appl. Immunol.,* **26,** 158 (1965).
309. R. Arnon and H. Neurath, *Immunochemistry,* **7,** 241 (1970).
310. K. C. Robbins and L. Summaria, *Immunochemistry,* **3,** 29 (1966).
311. P. O. Ganrot and J. E. Niléhn, *Scand. J. Clin. Lab. Invest.,* **24,** 15 (1969).
312. P. O. Ganrot and J. Stenflo, *Scand. J. Clin. Lab. Invest.,* **26,** 161 (1970).
313. P. O. Ganrot and J.-E. Niléhn, *Scand. J. Clin. Lab. Invest.,* **28,** 245 (1971).
314. J.-E. Niléhn and P. O. Ganrot, *Scand. J. Clin. Lab. Invest.,* **22,** 17 (1968).
315. J. Stenflo, *J. Biol. Chem.,* **248,** 6325 (1973).
316. J. Stenflo and P.-O. Ganrot, *J. Biol. Chem.,* **247,** 8160 (1972).
317. J. Stenflo, *J. Biol. Chem.,* **247,** 8167 (1972).
318. G. L. Nelsestuen and J. W. Suttie, *J. Biol. Chem.,* **247,** 8176 (1972).
319. I. Björk and J. Stenflo, *FEBS Lett.,* **32,** 343 (1973).
320. E. E. Rickli, *Immunochemistry,* **12,** 629 (1975).
321. K. Ohlsson, *Clin. Chim. Acta,* **26,** 131 (1969).
322. R. B. Andersen, J. Gormsen, and P. H. Petersen, *Scand. J. Rheumatol.,* **1,** 75 (1972).
323. M. Saint-Paul, P. Rebeyrotte, L. Dérobert, J. Peillet, and J.-P. Labbé, *Med. Leg. Domm. Corp. (Paris),* **5,** 68 (1971).
324. M. Saint-Paul, P. Rebeyrotte, L. Dérobert, J. Peillet, and J. P. Labbé *Beitrr. Gericht. Med.,* **30,** 376 (1973).
325. B. Weeke, *Protides Biol. Fluids,* **19,** 547 (1972).
326. C.-B. Laurell, *Scand. J. Clin. Lab. Invest.,* **17,** 297 (1965).
327. P. Rebeyrotte and J.-P. Labbé, *Bordeaux Medical,* **7-8,** 2105 (1971).
328. M. R. J. Salton and M. T. Schor, *Biochem. Biophys. Res. Commun.,* **49,** 350 (1972).

329. J. M. Andreu and E. Muñoz, *Biochim. Biophys. Acta,* 387, 228 (1975).

329a. M. Huberman, M. R. J. Salton, and P. Owen, Unpublished results (1975).

329b. P. Owen, M. Huberman, and M. R. J. Salton, Unpublished results (1975).

329c. M. R. J. Salton and P. Owen, Unpublished results (1975).

330. E. Muñoz, M. R. J. Salton, M. H. Ng, and M. T. Schor, *Eur. J. Biochem.,* 7, 490 (1969).

330a. M. R. J. Salton, W. Chaovapong, and P. Owen, Unpublished results (1975).

331. M. Lastras and E. Muñoz, *J. Bacteriol.,* 119, 593 (1974).

331a. M. Huberman and M. R. J. Salton, Unpublished results (1975).

332. C.-B. Laurell, *Protides Biol. Fluids,* 14, 499 (1967).

333. H. J. Firestone and S. B. Aronson, *Amer. J. Clin. Pathol.,* 52, 615 (1969).

334. W. Von Stephan and U. Frahm, *Z. Klin. Chem. Klin. Biochem.,* 8, 391 (1970).

335. B. Weeke and P. A. Krasilnikoff, *Protides Biol. Fluids,* 18, 173 (1971).

336. E. Schuller, B. Allinquant, M. Garcia, M. Lefèvre, P. Moreno, and L. Tompé, *Clin. Chim. Acta,* 33, 5 (1971).

337. A. G. M. Clarke, T. Freeman, and W. Pryse-Phillips, *J. Obstet. Gynaecol. Brit. Commonwealth,* 78, 105 (1971).

338. J. Lyngbye and J. Krøll, *Clin. Chem.,* 17, 495 (1971).

339. P. O. Ganrot, *Scand. J. Clin. Lab. Invest.* 29(Suppl. 124), 83 (1972).

340. B. Weeke and P. A. Krasilnikoff, *Acta Med. Scand.,* 192, 149 (1972).

341. J. M. B. Versey, L. Slater, and J. R. Hobbs, *J. Immunol. Methods,* 3, 63 (1973).

342. R. F. Ritchie, C. A. Alper, J. Groves, N. Pearson, and C. Larson, *Amer. J. Clin. Pathol.,* 59, 151 (1973).

343. P. A. Krasilnikoff and B. Weeke, *Protides Biol. Fluids,* 18, 169 (1971).

344. B. Abrams, *Clin. Sci.,* 40, 67 (1971).

345. C.-B. Laurell, *Scand. J. Clin. Lab. Invest.,* 21, 136 (1968).

346. A. H. Amin, H. G. M. Clarke, T. Freeman, I. M. Murray-Lyon, P. M. Smith, and R. Williams, *Clin. Sci.,* 38, 613 (1970).

347. K.-F. Aronsen, G. Ekelund, C.-O. Kindmark, and C.-B. Laurell, *Scand. J. Clin. Lab. Invest.,* 29,(Suppl. 124), 127 (1972).

348. E. Bock, *Scand. J. Immunol.,* 2(Suppl. 1), 111 (1973).

349. E. Bock and O. J. Rafaelsen, *Danish Med. Bull.,* 21, 93 (1974).

350. E. Bock, B. Weeke, and O. J. Rafaelsen, *J. Psychiat. Res.,* 9, 1 (1971).

351. H. G. M. Clarke, T. Freeman, and W. E. M. Pryse-Phillips, *Brit. J. Exp. Pathol.,* 51, 441 (1970).

352. H. G. M. Clarke, T. Freeman, and W. E. M. Pryse-Phillips, *Clin. Chim. Acta,* 30, 65 (1970).

353. H. G. M. Clarke, T. Freeman, and W. Pryse-Phillips, *J. Neurol. Neurosurg. Psychiat.,* 33, 694 (1970).

354. H. G. M. Clarke, T. Freeman, and W. Pryse-Phillips, *Clin. Sci.,* 40, 337 (1971).

355. H. G. M. Clarke, T. Freeman, R. Hickman, and W. E. M. Pryse-Phillips, *Thorax,* 25, 423 (1970).

356. H. G. M. Clarke, D. B. Grant, and D. Putman, *Arch. Disease Childhood,* 48, 608 (1973).

357. B. Forkman, P. O. Ganrot, G. Gennser, and G. Rannevik, *Scand. J. Clin. Lab. Invest.,* 29(Suppl. 124), 89 (1972).

358. J. Hällèn and C.-B. Laurell, *Scand. J. Clin. Lab. Invest.,* 29(Suppl. 124), 97 (1972).

359. S. Jeppson, C.-B. Laurell, and G. Rannevik, *Scand. J. Clin. Lab. Invest.,* 31, 33 (1973).

360. B. G. Johannsson, C.-O. Kindmark, E. Y. Trell, and F. A. Wollheim, *Scand. J. Clin. Lab. Invest.,* 29(Suppl. 124), 117 (1972).

361. C.-O. Kindmark and C.-B. Laurell, *Scand. J. Clin. Lab. Invest.,* 29(Suppl. 124), 105 (1972).

362. I. M. Murray-Lyon, H. G. M. Clarke, K. McPherson, and R. Williams, *Clin Chim. Acta,* 39, 215 (1972).

363. I. M. Murray-Lyon and R. Williams, *Clin. Chim. Acta,* 51, 303 (1974).

364. M. A. Pizzolato, *Clin. Chim. Acta,* 45, 207 (1973).

365. W. Von Stephan and U. Frahm, *Z. Klin. Chem. Klin. Biochem.,* 8, 469 (1970).

366. B. Weeke and S. Jarnum, *Gut,* 12, 297 (1971).

367. B. Weeke, E. Weeke, and G. Bendixen, *Acta Med. Scand.,* 189, 113 (1971).

368. B. Weeke, E. Weeke, and G. Bendixen, *Acta Med. Scand.,* **189,** 119 (1971).
369. U. Rovnskov, *Scand. J. Urol. Nephrol.,* **8,** 37 (1974).
370. C. A. Miller and A. W. Anderson, *Infect. Immunity,* **4,** 126 (1971).
371. P. J. Svendsen and N. H. Axelsen, *J. Immunol. Methods,* **1,** 169 (1972).
372. N. H. Axelsen and C. H. Kirkpatrick, *J. Immunol. Methods,* **2,** 245 (1973).
373. N. H. Axelsen, C. H. Kirkpatrick, and R. H. Buckley, *Clin. Exp. Immunol.,* **17,** 385 (1974).
374. N. Høiby, *Acta Pathol. Microbiol. Scand. Sect. B,* **82,** 551 (1974).
375. N. Høiby and L. Mathieson, *Acta Pathol. Microbiol. Scand. Sect. B,* **82,** 559 (1974).
376. N. Høiby and A. Wiik, *Scand. J. Resp. Diseases,* **56,** 38 (1975).
377. N. H. Axelsen, M. Harboe, O. Closs, and T. Godal, *Infect. Immunity,* **9,** 952 (1974).
378. B. Weeke, H. Lowenstein, and L. Nielsen, *Acta Allergol.,* **29,** 402 (1974).
379. G. Weinbaum and R. Markman, *Biochim. Biophys. Acta,* **124,** 207 (1966).
380. E. Alpert, J. W. Drysdale, and K. J. Isselbacher, *Ann. N.Y. Acad. Sci.,* **209,** 387 (1973).
381. R. L. Myerowitz, Z. T. Handzel, and J. B. Robbins, *Clin. Chim. Acta,* **39,** 307 (1972).
382. L. Røtbol, *Clin. Chim. Acta,* **29,** 101 (1970).
383. J. Lebas, LKB Application Note, LKB 2117 Multiphor No. 152, June (1974).
384. J. Lebas, A. Hayem, and J.-P. Martin, *C. R. Acad. Sci. Paris Ser. D,* **278,** 2359 (1974).
385. S. Harada, *Clin. Chim. Acta,* **63,** 275 (1975).
386. J. Krøll, *Protides Biol. Fluids,* **19,** 529 (1972).
387. C.-B. Laurell, in C. Mittman, Ed., *Pulmonary Emphysema and Proteolysis,* Academic Press, New York, 1972, p. 349.
388. S. Walbaum, J. Biquet, T. Vaucelle, and P. Tran van Ky, *Bull. Soc. Franc. Mycol. Med.,* **2,** 3 (1974).
389. S. Walbaum, T. Vaucelle, and J. Biquet, *Pathol. Biol.,* **21,** 355 (1973).
390. O. J. Bjerrum, S. Bhakdi, H. Knüfermann, and T. C. Bøg-Hansen, *Biochim. Biophys. Acta,* **373,** 44 (1974).
391. A. J. Crowle, A. A. Atkins, and G. J. Revis, *J. Immunol. Methods,* **4,** 173 (1974).
392. C. E. Thirkhill and G. E. Kenny, *Infect. Immunity,* **10,** 624 (1974).
393. W. Groc and A. Harms, *J. Immunol. Methods,* **8,** 85 (1975).
394. J. R. Green, M. J. Dunn, and A. H. Maddy, *Biochim. Biophys. Acta,* **382,** 457 (1975).
395. J. J. Volpe and P. R. Vagelos, *Annu. Rev. Biochem.,* **42,** 21 (1973).
396. C.-B. Laurell and E. Thullin, *Scand. J. Immunol.,* **4**(Suppl. 2), 7 (1975).
397. H. G. M. Clarke and T. Freeman, *Protides Biol. Fluids,* **14,** 503 (1967).
398. T. Wadström and C. J. Smyth, in P. G. Righetti, Ed., *Progress in Isoelectric Focusing and Isotachophoresis,* North Holland Publishing Company, Amsterdam, 1975, p. 149.
399. J. N. Baptist, C. R. Shaw, and M. Mandel, *J. Bacteriol.,* **108,** 799 (1971).
400. D. C. Cann, G. Hobbs, and J. M. Shewan, in B. M. Gibbs and F. A. Skinner, Eds., *Identification Methods for Microbiologists,* Part A, Academic Press, London, 1966, p. 97.
401. K.-E. Johansson and S. Hjertén, *J. Mol. Biol.,* **86,** 341 (1974).
402. C.-H. Brogren, G. Peltre, T. C. Bøg-Hansen and Å. Hansen, *Protides Biol. Fluids,* **24,** in press (1976).
403. C.-H. Brogren, in B. J. Radola and D. Graesslin, Eds., *Electrofocusing and Isotachophoresis,* Walter de Gruyter, Berlin and New York, in press (1976).

CHAPTER 7

THE ANTIGENICITY OF CYTOCHROME *C*

G. J. URBANSKI AND E. MARGOLIASH

1. INTRODUCTION

Ever since globular proteins were shown to be powerful antigens in the late nineteenth century, there have been numerous studies aimed at determining what structures are responsible for such effects and for their remarkable specificity. Among these are the classic studies of Heidelberger and Landsteiner (1), who in 1923 observed precipitating antibodies that could distinguish among the hemoglobins of various species. Antibodies elicited by the native protein did not react with the denatured protein (2), a phenomenon now attributed to antigenic determinants, the structures of which depend upon the native spatial conformation of the protein (3–5). These determinants have been termed conformational, to operationally distinguish them from a second class of determinants, which arise from amino acid sequences that exist in a random coil form (3) and are termed sequential. The antibodies to sequential determinants react with relatively small peptide fragments that can assume many conformations in solution, whereas antibodies to conformational determinants react only with the immunogen in its original conformation, the maintenance of which commonly requires a large proportion, if not all, of the protein. It should be stressed that these definitions are, in fact, based on quantitative rather than on real qualitative differences, because even antibodies to conformational determinants can be inhibited from reacting with the native protein by large molar excesses of the corresponding peptide (6), a minute fraction of which may be in the native protein conformation.

Another problem with this simple distinction between sequential and conformational determinants is that it does not take into account that there are two fundamentally different ways in which immunologically recognizable alterations in the structure of a native protein can occur. Evolutionary changes in the amino acid sequences in a homologous set of proteins may cause changes in the spatial arrangement of the polypeptide backbone, resulting in the formation of antigenic determinant areas on the protein recognizable by the immune system of a particular host. Since the position in space of the polypeptide backbone is what is commonly and properly termed the conformation of a protein, we would like to restrict the term conformational determinant to the cases in which a change in backbone location constitutes or contributes to the antigenic site or sites. Such variations are more likely to occur when the amino acid substitution occurs at an internal residue position; most globular proteins are so tightly packed that any internal change is likely to displace the backbone. Amino acid substitutions at the surface can also conceivably cause changes in backbone structure; in such cases, the resultant antigenic determinants may be located both at the site of the amino acid substitution and/or at remote portions of the protein.

The more common situation is that surface substitutions cause no change in the backbone. These replacements, nevertheless, result in effective antigenic determinants. In such cases, the immune system of a particular host recognizes only the local change in the surface of the protein. Such determinants should be called topographic, a term that is more descriptive of the actual situation. Topographic determinants are common for the hemoglobins (7) and cytochromes *c* and are likely to be important in other groups of globular proteins.

Cytochrome *c*, a heme protein of the mitochondrial respiratory chain, consists of a single-chain polypeptide of more than 100 residues. It is found in all eukaryotes; pro-

Supported by Grants AI-12001, GM-19121, and HL-1119 from the National Institutes of Health.

karyotes have related proteins of different functional characteristics (8a,8b). The amino acid sequences of more than 75 of these proteins from different eukaryotes have been determined (9,10). Furthermore, the spatial conformations of the cytochromes *c* from the horse, tuna, and bonito (8b,11) have been found to be identical and are certainly the same for all eukaryotic cytochromes *c*. This group of proteins represents an excellent system with which to study the effects of amino acid residue variations on the antigenicity of a globular protein, independently of changes in conformation.

2. IMMUNOGENICITY OF CYTOCHROME *c*

Interest in the immunogenicity of cytochrome *c* was sparked in the 1940s because of its possible, though questionable, therapeutic value. Proger and Dekaneas (12) reported in 1946 that injection of cow cytochrome *c* in man was beneficial for the treatment of a variety of conditions that involve tissue anoxia and that the injected protein persisted in the serum and various tissues. At the same time, they also reported that the cytochrome *c* used, a Keilin and Hartree type of preparation of cow cytochrome *c* that contains 0.34% iron (13), "appeared to be non-antigenic." No evidence to support this assertion was given. Roth et al. (14), on the other hand, reported in 1949 that commercial preparations of cytochrome *c* were, indeed, immunogenic; they were capable of producing anaphylaxis in sensitized guinea pigs.

To resolve this conflict, Becker and Munoz (15) in 1949 conducted a detailed study of the immunogenicity of available cytochrome *c* preparations. The purity of their materials was determined by measuring the iron content. At this time, a cytochrome *c* preparation that contained 0.43% iron was considered to be entirely pure from the work of Theorell and Åkesson, which dates back to 1941 (16); pure cytochrome *c*, in fact, contains 0.45% iron (17–21). Becker and Munoz (15) found that two preparations of cow cytochrome *c*, which contained 0.33 and 0.35% of iron, respectively, were immunogenic in rabbits; the antibodies were detected by the ring test (22). A preparation of hog cytochrome *c* (0.38% iron) also produced an immune response. However, a preparation of horse cytochrome *c* of 0.43% iron did not. In all of these cases, the antigens were presented as either alum precipitates or in complete Freund's adjuvant (CFA), by use of a total dose of 4–50 mg over 1–4 weeks. The gel diffusion technique of Oudin (23,24) was used to show that these antisera contained a single antibody-antigen system, and an ingenious assay was developed to determine whether the cytochrome *c* itself was the immunogenic component of these preparations. Solutions of the antigens were precipitated by successive additions of antibody, and the cytochrome *c* concentration of each supernatant fluid was determined. It was found that the antibodies produced against the antigenically active preparations were directed against colorless protein impurities and not the cytochrome *c* itself. At about the same time, Beinert (25) showed that, contrary to the earlier report of Proger and Dekaneas (12), ^{55}Fe-labeled cytochrome *c* injected intravenously is rapidly and essentially quantitatively eliminated through the kidneys or broken down without mixing with intracellular protein, which makes it doubtful that any specific systemic physiologic or pharmacologic effects could have been obtained. Despite this evidence, studies on the presumed clinical effects of cytochrome *c* continued to appear at irregular intervals. Work on the immunogenicity of the protein also continued, and the reports were once more in conflict.

Storch et al. (26) in 1964 found that purified preparations (0.44% iron) of horse cytochrome c were nonimmunogenic in rabbits and guinea pigs by several criteria. Rabbits were given four intraperitoneal injections of antigen per week for 3 weeks. Each injection contained from 5 to 100 mg of cytochrome c per kilogram of body weight in CFA. No antibody was detected in these sera 8–15 days after the last injection by either ring or Ouchterlony tests (22,27). They also attempted to sensitize guinea pigs by use of one of three successive intraperitoneal or subcutaneous injections of from 20 to 100 mg of protein per kilogram of body weight per injection. Negative results were obtained by the Schultz-Dale test (22) with the guinea pig ileum. It is now clear that the antigen levels used in this study were far in excess of doses found to be most immunogenic, and the observed lack of respose was probably due to the induction of immunologic paralysis.

In 1966, Jonsson and Paléus (28) detected low levels of rabbit antibodies against chicken cytochrome c by the mixed hemabsorption test. With the same sera, the passive hemagglutination test was negative. A similar reaction was not obtained with cow cytochrome c, probably because of individual differences among rabbits rather than differences in immunogenicity of the cytochromes c, because only one rabbit was immunized against each antigen.

Reichlin et al. (29) and Watanabe et al. (30) obtained weak antibody responses in rabbits against native monomeric horse cytochrome c with a prolonged immunization routine. Reichlin et al. (29) gave two subcutaneous injections of 5 mg of protein in CFA 10 days apart, followed by weekly intravenous injections of 5 mg in saline. Watanabe et al. (30) obtained similar results with eight to 10 subcutaneous injections of 10 mg of cytochrome c in CFA every other week.

Reichlin et al. later obtained rabbit antisera against native human, *Macaca mulatta,* kangaroo, turkey, and tuna cytochromes c (31–33). They found that the antisera elicited by the human, monkey, and horse proteins precipitated and fixed complement with the native monomeric homologous cytochromes c, whereas those directed against kangaroo, turkey, and tuna cytochromes c precipitated or fixed complement only with polymers of the protein (32). They also reported that of the cytochromes c used, only the human protein was a consistent and potent immunogen when presented in the monomeric state (32). One explanation for this finding may be the fact that the human protein was prepared in small amounts from autopsy material, which would increase the probability that these presumably monomeric preparations were contaminated with small amounts of polymer, a form of the protein that is a much more effective immunogen (34). Tuna and turkey cytochromes c elicited stronger antibody responses, more promptly and in a higher precentage of animals, when acetylated bovine γ-globulin-conjugated antigen rather than the native protein was employed for immunization (32).

Okada et al. (35) and Watanabe et al. (30) reported that baker's yeast (*Saccharomyces oviformis*) cytochrome c could evoke a considerable immune response in both rabbits and mice. Watanabe et al. (30) injected rabbits subcutaneously with 10 mg of the baker's yeast protein in CFA for a total of three injections. The protein was presented both as the disulfide dimer and as the carboxymethylcysteine monomer. These experiments were necessary because the baker's yeast protein is the only cytochrome c known to contain a single free cysteine sulfhydryl group (residue 102) and to spontaneously form dimers (36). A strong response was obtained in rabbits against the disulfide dimer (700 μg of antibody/ml), as determined by the quantitative precipitin test. Less antibody was formed against the carboxymethylated stable monomer, but both responses were qualitatively the same by the Ouchterlony test. Mouse antibodies against the dimer were detected by passive cutaneous anaphylaxis. Watanabe et al. (30) and Margoliash et al. (32) demonstrated that the rabbit anticytochrome *c* activity was limited to the IgG immunoglobins. The mouse anticytochrome *c* activity was also shown to reside in the γ-1 and γ-2 immunoglobulins (30).

The most useful advance came when Reichlin et al. (37) obtained strong and consistent immune responses in rabbits against a large number of cytochromes *c* with polymers of the protein cross-linked with glutaraldehyde. This observation correlates well with the data available for other proteins, such as human globulin, which show that aggregated forms of the protein are highly immunogenic (38), whereas purified monomer preparations readily induce tolerance rather than yielding antibodies (39,40). After short courses of immunization, which lasted 1 month, polymers of human, monkey, horse, kangaroo, turkey, and tuna cytochromes *c* elicited amounts of antibodies comparable to those elicited by good soluble antigens, such as serum albumins and globulins. With a comparable immunization schedule, all monomeric cytochromes *c* except the human protein, as noted above, did not yield detectable antibodies. Low levels of antibody were obtained with monomers in a small proportion of rabbits only if the immunization was continued for 3 or more months. A comparison of the available precipitin data for monomers and polymers is given in Table 1.

All of the approximately 40 antisera elicited by polymeric cytochromes *c* contained no antibodies that reacted specifically with polymers without also binding to the corresponding monomer. Furthermore, for the antisera that did not precipitate or fix complement with the corresponding monomeric cytochrome *c* (tuna, turkey), it was shown that the homologous monomers inhibited more than 95% of the precipitation and complement fixation reactions of the polymers, regardless of whether the antisera were produced against the monomeric or polymeric form of the antigen. This inhibition was 50% when the monomer and polymer were present in equal amounts, demonstrating

Table 1. Immunogenicity of Monomers and Polymers of Cytochromes *c* from Several Eukaryotes in Rabbits

Cytochrome *c*	Form	Time course of immunization (months)	Average precipitin content of antisera[a] (μg of AbN/ml)	Reference
M. mulatta	Monomer	≥3	20	37
Kangaroo[b]	Monomer	≥3	35	37
Horse	Monomer	≥3	40	30, 37
Human	Monomer	3	70	37
Baker's yeast	Dimer	1	92	30
Rabbit	Polymer	1	32	37
Mouse	Polymer	1	41	41, 42
Guanaco	Polymer	1	58	41, 42
Kangaroo	Polymer	1	104	37
M. mulatta	Polymer	1	168	37
Horse	Polymer	1	219	37, 42
Human	Polymer	1	221	37
Tuna	Polymer	1	221	37

[a] Precipitable antibody determined by use of the same physical form of the antigen as that injected. The values listed are averages given in order of increasing antibody content, not counting nonresponding animals.

[b] Precipitable antibody determined by use of polymer.

that the binding affinity of the antibodies to the two forms of the antigen were the same. A small immune response was also elicited in rabbits against glutaraldehyde polymers of rabbit cytochrome *c* (37) (see Table 1). These antibodies were also shown to be identical in their reactions with both the polymeric and monomeric forms of the antigen.

Reichlin and Turk (43) also induced delayed hypersensitivity against cytochrome *c* in guinea pigs. The animals were skin tested 2 weeks after single footpad injections of 1.0 mg of horse cytochrome *c* in CFA. The responses displayed the same pattern of cross reactivity to other cytochromes *c* that is displayed by the humoral response. However, in contrast to the humoral response, the cellular phenomenon differentiates between the monomeric and polymeric forms of the antigen (43a). Guinea pigs primed with polymers display a stronger response to polymers than monomers, while the opposite is the case for animals primed with monomers. One explanation for this difference may be that guinea pigs are able to recognize glutaraldehyde on the surface of the cytochrome *c* and rabbits are not. This explanation seems unlikely, because, as will be discussed later, the rabbit immune system is capable of detecting very small changes on the surface of the proteins. Also, preliminary studies of the kinetics of the polymerization reaction suggest that all of the cross-linking reagent moieties may be buried inside the polymer complex (44). Thus, it is likely that the differences lie at the cellular level, with the T cell (thymus-derived lymphocyte) differentiating between the single antigenic determinants of the monomer and the clustered antigenic determinants of the polymer.

Concurrently with the immunologic studies on cytochrome *c*, massive amounts of data were being accumulated on the primary and tertiary structure of this group of proteins. It is this detailed knowledge of structure that has made cytochrome *c* such a useful model globular protein antigen.

3. STRUCTURE OF CYTOCHROME *c*

Following the determination of the amino acid sequence of horse cytochrome *c*, which was completed in 1960 (45–47), the proteins from a large number of species have enjoyed the attention of several groups of investigators. Their primary structures are relatively simple to establish, because their single polypeptide chains are only a few residues longer than 100 (Fig. 1); they are easily prepared and are ubiquitous among eukaryotes. Cytochrome *c* has been extensively used as a prime model for the development of the statistical techniques needed to extract evolutionary information from protein amino acid sequences (48–51), encouraging the examination of an increasingly wider range of taxonomic groups. As a result of this activity, the primary structures of the cytochromes *c* of more than 75 eukaryotic species are now known (9,10).

The tabulation of these amino acid sequences (Fig. 2) gives an overwhelming impression of variability. Counting from the invariant glycine at residue 1, there are only 21 unvaried positions, while as many as 42 positions can be accommodated by five or more different amino acids. However, this does not mean that any combination of the residues that occupy the various positions in the different cytochromes *c* of known structure would constitute a cytochrome *c*. As shown by Fitch and Markowitz (53), the number of positions that are variable in any one cytochrome *c* is restricted to the "covarions," estimated at about 10 residues, as compared to the 83 that are known to have varied among eukaryotic cytochromes *c* (9,10). These restrictions obviously result from the structural and functional requirements of the protein, and as long as these

*Gly - Asp-Val- Glu - Lys - Gly - Lys-Lys - Ile - Phe-Val-Gln- Lys - Cys-Ala
              5   10   15

HEME ⌐
Gln- Cys-His-Thr - Val - Asp-Lys-Gly - Gly - Lys-His - Lys-Thr - Gly-Pro
              20   25   30

Asn-Leu-His - Gly - Leu-Phe-Gly-Arg-Lys-Thr-Gly- Gln- Ala - Val-Gly
              35   40   45

Phe-Ser - Tyr - Thr-Asp-Ala-Asn-Lys -Asn-Lys-Gly -Ile - Thr -Trp-Gly
              50   55   60

Glu - Asp-Thr-Leu-Met- Glu - Tyr-Leu-Glu-Asn-Pro-Lys-Lys -Tyr-Ile
              65   70   75

Pro-Gly - Thr-Lys-Met-Ile-Phe-Ala-Gly -Ile - Lys-Lys-Lys-Asp-Glu
              80   85   90

Arg-Ala-Asp-Lys -Ile - Ala-Tyr-Leu-Lys-Lys-Ala-Thr-Asn-Glu-COOH
              95   100   104

Fig. 1. The amino acid sequence of rabbit cytochrome c according to Needleman and Margoliash (52). The asterisk indicates that the amino terminal glycine residue is N-acetylated. The bars from "HEME" to cysteinyl residues 14 and 17 represent the thioether bonds between the prosthetic group and the polypeptide chain.

requirements are not known in detail, it is not possible to say which sequence would actually represent a cytochrome c and which would not.

Nevertheless, from the point of view of the use of cytochrome c as a model globular protein antigen, the situation is particularly favorable. Indeed, there is little doubt that the conformation of the polypeptide backbone of all eukaryotic cytochromes c is the same (8b,49). This conclusion probably represents the most important single contribution of x-ray crystallography to the immunology of cytochrome c, because it signifies that differences in amino acid sequence yield only changes in local topography, as defined in Sec. 1 above, and that these changes can be correlated in a simple fashion with the products of the immune response.

To date, spatial structures have been determined by x-ray crystallography for horse, bonito, and tuna cytochromes c (8b), eukaryotic proteins that differ at a maximum by 19 residues, and for the C-type cytochromes of two prokaryotes, cytochrome c_2 of *Rhodospirillum rubrum* (58,59) and cytochrome c_{550} from *Paracoccus denitrificans* (60,61). The prokaryote proteins differ from those of the eukaryotes by two thirds or more of their amino acid sequences. All of these proteins maintain precisely the same arrangement of their polypeptide backbones in space, the so-called cytochrome c fold (8b,49) (see Figs. 3a–3d). Where additional segments of sequence are present over and beyond the simple cytochrome fold, these segments appear as external appendages; where there are fewer residues, a bend is tightened to make the peptide chain regain its original position in as short a segment as possible. Clearly, by far the most conservative character of the protein is its spatial arrangement. Once the evolutionary process had resulted in a molecular machine that satisfied the myriad of biologic demands of synthesis, stability, function, and self-assembly with other components of its organelle niche, and so on, the successful product has been most cautiously preserved and used in a large variety of

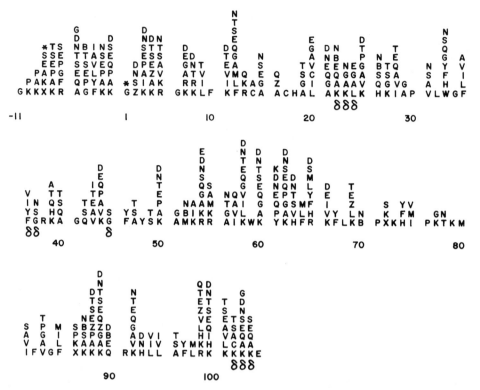

Fig. 2. Composite amino acid sequence of eukaryotic cytochromes c. The glutamic acid at position 105 is found in the ginkgo protein, the only one to extend that far. The asterisk at position −8 indicates that this residue is N-acetylated in the higher plant cytochromes c, and the asterisk at position 1 similarly denotes the acetyl present in all chordate cytochromes c. The cytochromes c tabulated are those of man, chimpanzee, rhesus monkey (*Macaca mulatta*), horse, donkey, cow, pig, sheep, dog, elephant, seal, rabbit, California gray whale, camel, great gray kangaroo, bat, mouse, chicken, turkey, pigeon, penguin,* duck, emu, ostrich,* rattlesnake, snapping turtle, bullfrog, tuna, bonito, carp, dogfish, Pacific lamprey, snail (*Helix aspersa*),* moth (*Samia cynthia*), tobacco hornworn moth, fruit fly (*Drosophila melanogaster*), screwworm fly, fly (*Ceratatis*) (54), brandling worm (*Eisenia foetida*) (55), fungus (*Neurospora*), baker's yeast (*iso*-1 and *iso*-2 cytochromes c), *Debaromyces, Candida krusei, Ustilago sphaerogena, Humicola,* the higher plants wheat, mungbean, sunflower, sesame, castor, cauliflower, buckwheat, pumpkin, *Abutilon,* cotton, ginkgo, elder,* nighe,* tomato,* spinach,* hemp,* leek,* parsnip,* potato,* love-in-the-mist,* nasturtium, box elder,* maize,* Arum,* algae (*Enteromorpha,* and *Polytoma* (56), and the protists *Physarum, Euglena, Crithidia oncopelti,* and *Crithidia fasciculata* (57), and *Tetrahymena.* References to the structures of these proteins are listed in Ref. 49, except those indicated by the asterisk, which, appear in Ref. 9 or those with specific references given.

The following single-letter code is used in this Figure: A, alanine; B, aspartic acid or asparagine; C, cysteine; D, aspartic acid; E, glutamic acid; F, phenylalanine; G, glycine; H, histidine; I, isoleucine, K, lysine; L, leucine; M, methionine; N, asparagine; P, proline; Q, glutamine; R, arginine; S, serine; T, threonine; V, valine; W, tryptophan; X, trimethyllysine; Y, tyrosine; Z, glutamic acid or glutamine; δ, deleted. The δ marks a gap that has been introduced in some proteins to maximize similarity.

permutations that do not impair its fundamental molecular arrangement. Like all proteins, cytochrome *c* is subject to the following three classes of constraints that govern its amino acid sequence and spatial conformation: structural constraints to meet the requirements of proper folding and interior close packing of the globular molecule; biochemical constraints that are set by the need to perform its function of appropriately interacting with its respiratory chain neighbors, transferring electrons, and serving the related control functions (11,49,62,63); and genetic constraints that provide for any cytochrome *c* to be an evolutionary descendant from its ancestral form via the allowable genetic and evolutionary manipulations. Here again, as long as structure-function relationships are not understood in their complete detail, it is impossible to know which structural feature relates to which type of constraint, other than in the primitive form of contrasting the interior with the exterior of the molecule.

To summarize, cytochromes *c* that differ as much as the proteins from eukaryotes and prokaryotes maintain the same spatial conformation of their polypeptide backbones, even though their amino acid sequences vary by more than two thirds of the polypeptide and the details of their functions are entirely different. This makes it certain that for eukaryotic cytochromes *c* that differ by a few residues and are functionally close to indistinguishable, the spatial conformations of the polypeptide backbone are precisely the same. This argument remains valid whether or not one considers electron density maps of proteins produced by present x-ray crystallographic techniques sufficiently sensitive to detect differences in structure that may be detected by the immune system. Because no differences are detected with massive changes in amino acid sequence, the probability that significant structural changes accompany minor residue substitutions is very small. When such minor sequence differences result in an immune response, we must therefore be dealing with topographic immunogenic determinants and not with conformational determinants, as defined in Sec. 1.

This conclusion is verified by the immunologic behavior of the cytochrome *c* determinants, detailed in the following sections. Indeed, the products of responses to local changes in the surface topography of the molecule can be simply analyzed in terms of these changes, any one being independent of any others present in the same molecule. Thus, antibodies raised against one such determinant will react identically with the protein used as the immunogen as with any other cytochrome *c* that carries the same determinant, irrespective of what other topographic determinants are available on the cross-reacting protein. The fact that every cytochrome *c* immunogenic determinant so far examined behaves in this fashion makes the idea that the immune response is due to a conformational change induced by the amino acid substitution and possibly at a distance from it (6) essentially untenable in the case of cytochrome *c*.

The folding pattern typical of cytochrome *c* can be easily followed in Figs. 3a–3d, particularly if the reader manages to grasp the three-dimensional effect of the stereoscopic diagram. This is most readily accomplished by holding the Figure at a distance that helps the right eye see the right diagram and the left eye see the left diagram. When the eyes look through the page rather than focus on it, the two images fuse, and a third brighter three-dimensional figure will appear between the other two. Good lighting is helpful. The amino terminal segment (from the carboxyl group of residue 2 to the amino group of residue 14) is an α-helix that goes from the top back to the top front of the molecule. This segment is followed by that which contains cysteinyl residues 14 and 17, which bind the polypeptide covalently to the heme prosthetic group via thioether bonds to the α-carbons of the vinyl side chains of pyrrole rings I and II. The chains turn back

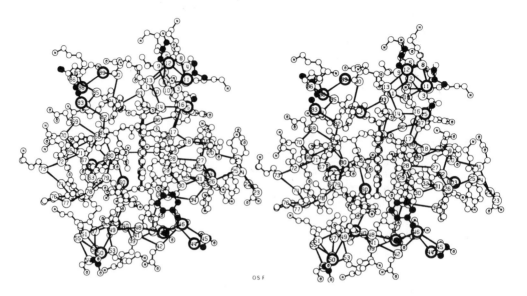

Fig. 3a. Stereoscopic diagram of tuna cytochrome c in the ferric form as viewed from the front, according to Swanson et al. (64), from an electron density map at a resolution of 2.0 Å. The larger circles are the α-carbon atoms, while the smaller circles represent the side-chain atoms. The heme is the square structure inserted in the front of the molecule. It is seen edge on. The heavier circles around some of the α carbons and the black side chain atoms indicate the residues that vary between the cytochromes c listed in Fig. 4 (rabbit, mouse, guanaco, cow, dog, horse, and human). The amino acid sequence of tuna cytochrome c (65,66) is: *G-D-V-A-K-G-K-K-T-F-V-Q-K-C-A,-Q-C-H-T-V-E-N-G-G-K-H-K-V-G-P-N-L-W-G-L-F-G-R-K-T-G-Q-A-E-G-Y-S-Y-T-D-A-N-K-S-K-G-I-V-W-N-N-D-T-L-M-E-Y-L-E-N-P-K-K-Y-I-P-G-T-K-M-I-F-A-G-I-K-K-K-G-E-R-Q-D-L-V-A-Y-L-K-S-A-T-S-δ. See Fig. 2 for a listing of the single-letter code.

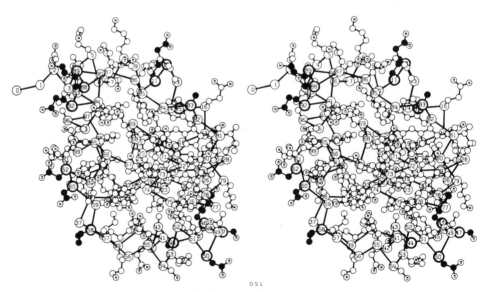

Fig. 3b. Stereoscopic diagram of tuna cytochrome c in the ferric form as viewed from the left side. The front of the molecule is toward the right of the Figure. The heme is the square structure, seen from one side, located at the front half of the molecule near its center. Other details are as given in the legend to Fig. 3a.

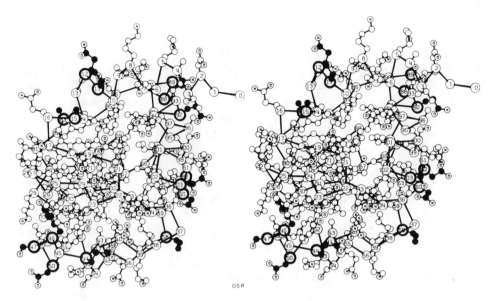

Fig. 3c. Stereoscopic diagram of tuna cytochrome c in the ferric form as viewed from the right side. The front of the molecule is to the left of the Figure. The heme is the square structure, seen from one side, located at the front half of the molecular near its center. Other details are as given in the legend to Fig. 3a.

sharply from the anterior thioether bonded cysteine (residue 17) to histidyl residue 18, the uniquely invariant histidine that provides the imidazole group coordinated to the central heme iron atom from the left side of the molecule. The chains go backward and forward twice to complete the left half of the structure and the bottom of the heme crevice. There are short α-helical segments at residues 49–55, 60–70, and 70–75. The

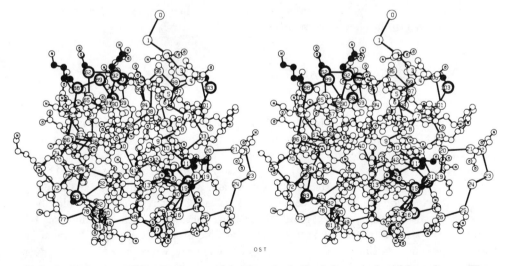

Fig. 3d. Stereoscopic diagram of tuna cytochrome c in the ferric form as viewed from the top. The front of the molecule is toward the bottom with the back toward the top. Other details are as given in the legend to Fig. 3a.

213

chain then follows a path on the right side from the middle back to the bottom front, up to the middle center, at which residue 80 yields the methionine sulfur coordinated to the heme iron from the right, and then up and back to the top back right edge. This segment is followed by the carboxylterminal α-helix (residues 87–103), which runs nearly perpendicular to the heme plane, and to the amino terminal α-helix, across the back of the molecule from left to right, and ends in the last three residues of the chain.

The molecule is an excellent representative of the "oil drop" model of protein structure with the hydrophobic side chains mostly in the interior and the hydrophilic side chains on the outside. Any interior hydrophilic groups fulfill specific structural purposes, such as hydrogen bonding or providing the needed heme iron ligands. The square heme prosthetic group is nearly entirely enrobed in the hydrophobic interior of the protein; only the edge that contains pyrrole rings II and III appears at the center front of the molecule. The propionic acid side chains of pyrrole rings III and IV point down into the bottom of the heme crevice; pyrrole ring IV is at the back, with its side deep inside the protein and held down by three hydrogen bonds. This arrangement is in sharp contrast with that of hemoglobin and myoglobin, in which the propionyl side chains are in the outer aqueous environment. The heme, which is nearly entirely inside the protein, is in contact with a large proportion of all interior residues and, itself, therefore provides a major portion of the forces that direct the overall folding pattern of the polypeptide chain. Eighteen residues of the total of 104 are in contact with the heme. Like all internal residues, these residues tend to be evolutionarily conservative, whereas residues that show a large number of evolutionary transformations are external. All of the residues that constitute the immunogenic determinants so far examined in any detail have external side chains. The cytochromes *c* that have the rare internal residue substitutions, as compared to the proteins of experimental animals available for immunization, will be from species taxonomically far removed and therefore also carry a large number of external substitutions. In such cases, it is experimentally very difficult to analyze the immunologic influence, if any, of the internal substitution on the complex background presented by the immunologic responses to the many external determinants. So far, this analysis has not been attempted.

4.   IDENTIFICATION OF ANTIGENIC SITES ON CYTOCHROME *c*

As can be seen from a comparison of the amino acid sequences of several mammalian cytochromes *c* (Fig. 4) and the structure of the protein in space (Figs. 3a–3d), the two to nine amino acid sequence differences can be grouped into seven potentially antigenic regions with respect to mammalian hosts. The variant residues of region I are 11, 12, and 15, the side chains of which are all external and in close proximity, near the top of the heme crevice on the front of the protein (Fig. 3a). Other residues that could affect this region are invariant in the group of proteins considered, such as glutamine 16. Thus, their possible contributions to the immunologic properties of region I cannot be studied without artificial chemical modifications. The next two regions involve residues 44–50. Notwithstanding their closeness in the amino acid sequence, the side chain of residue 44 is pointing toward the right side at the bottom right of the protein, while the side chains of residues 47 and 50 are pointing straight forward at the bottom front of the protein. The side chain of residue 46 is aromatic and flat on the surface of the molecule

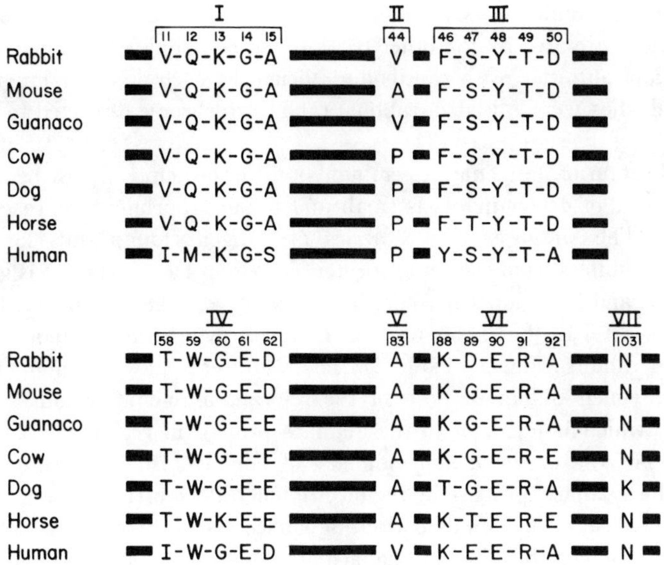

Fig. 4. Comparison of the amino acid sequences of some mammalian cytochromes *c*. The havy bars indicate portions of the sequences that are identical and are porportional in length to the number of residues omitted. The single-letter code used is as follows: A, alanine; D, aspartic acid; E, glutamic acid; F, phenylalanine; G, glycine; I, isoleucine; K, lysine; M, methionine; N, asparagine; P, proline; Q, glutamine; R, arginine; S, serine; T, threonine; V, valine; W, tryptophan; Y, tyrosine.

also near the bottom front. It is the sharp turn at residue 45, an invariant glycine, that divides this segment into two immunologic regions; residue 44 is in region II, and the remainder is in region III (Fig. 3a). Region IV is at the lower half of the back of the protein and contains residues 58, 60, and 62 with their side chains pointing backward (Figs. 3b and 3d). Region V contains residue 83 at the top left corner of the molecule. This residue is about 15 Å from the center of region I, and an antibody that binds at either site might interfere with binding at the other (Fig. 3a). Region VI contains residues 88, 89, and 92 at the back top left of the protein (Figs. 3b and 3d). The side chains of these residues are again external and in close proximity to each other. The last potentially antigenic area (region VII) contains residue 103 in the middle of the right edge of the back surface of the protein. The side chain of this residue can be seen projecting backward in Fig. 3c.

The first experiments, in which the cross reactivities of a large variety of different cytochromes *c* with rabbit antihuman and antihorse cytochrome *c* sera were determined, showed an adequate but not perfect correlation between the percentage cross reactivities and the extent of the amino acid sequence differences between the immunizing and cross-reacting proteins (31). These studies employed a modification of the Farr technique (67) and inhibition of complement fixation. With the antibody fluorescence quenching determination of the stoichiometry of the reactions by the procedure of Noble et al. (68), even closer correlations were observed. These stoichiometries are a measure of the number of independent antigenic sites, in terms of the number of Fab fragments capable of binding simultaneously to the cytochrome *c* molecule, whereas the percentage cross reactivities estimated by the other techniques are strongly dependent on the varying amounts and affinities of the antibodies elicited by each determinant. Even though all of the antibodies were raised in rabbits, rabbit cytochrome *c* reacted to considerable degrees with all of the sera tested (31).

The identification of antigenic determinants on the protein can best be accomplished when the number of determinants is small and the appropriate cross-reactive proteins are available. The simplest of this series of antigens found thus far is guanaco cytochrome *c*. It differs from the rabbit protein in region IV, where it has glutamic acid acid residue 62, and in region VI, where it carries a glycine at position 89. Both of these residues are aspartyl in the rabbit protein. Guanaco cytochrome *c* elicits two antibody populations in rabbits (Fig. 5, Table 2). These antisera show no binding with horse cytochrome *c*, which differs in both of the relevant antigenic regions, whereas cow cytochrome *c*, which is identical to the guanaco protein in region IV but different in region VI, yields the expected stoichiometry of one. Mouse cytochrome *c*, which is identical to the guanaco protein in region VI but differs in region IV by having an aspartic acid at position 62, binds both antiguanaco cytochrome *c* populations. This is probably due to the fact that the antiguanaco antibody populations elicited by region IV that contains a glutamic acid also bind to this region when it contains an aspartic acid. Indeed, these antiguanaco cytochrome *c* sera display a stoichiometry of one with the rabbit protein. When isolated, this population binds with a stoichiometry of one to rabbit, mouse, guanaco, and cow cytochromes *c*, all of which have either an aspartic or a glutamic acid at residue 62, but fails to bind the horse protein because of the interference

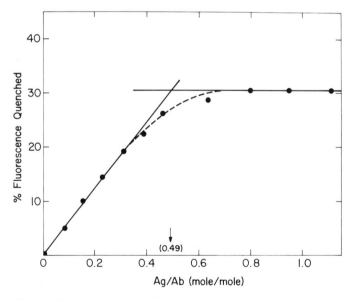

Fig. 5. Titration of rabbit antiguanaco cytochrome *c* Fab fragments (1.27 × 10⁻⁶ M) with guanaco cytochrome *c* (41). The percentage of the antibody fluorescence quenched is plotted against the molar ratio of cytochrome *c* to reactive Fab fragments (68). The stoichiometry of the reaction is defined by the reciprocal of the intersection of the initial slope and the line of maximal fluorescence quenched.

Table 2. Stoichiometries of the Reaction of Rabbit Antiguanaco Cytochrome *c* Antibodies with Various Cytochromes *c*[a]

Antibody	Cytochrome *c*			
	Guanaco	**Mouse**	**Cow**	**Rabbit**
Antiguanaco cytocrhome *c*	2.01 ± 0.06	1.88 ± 0.17	nd[b]	1.02 ± 0.04
Antiguanaco region IV	1.04 ± 0.02	1.06 ± 0.04	1.03	1.01 ± 0.06
Antiguanaco region VI	1.02 ± 0.02	1.01 ± 0.03	0[c]	0

[a] The titration of 2.0 ml of rabbit antiguanaco cytochrome *c* Fab fragments was performed at concentrations of about 10^{-6} M in borate-buffered saline (pH 8.6) with increasing amounts of the cytochromes listed, which ranged from 0.1 to 6.0 nmol. The values given represent the averages of several different animals and bleeding times ± the standard deviation from the mean.

[b] Not determined (42).

[c] No binding observed.

caused by the introduction of a lysyl side chain at residue 60 (Table 2, Fig. 4). This pattern demonstrates that of the two antibody populations in the antiguanaco cytochrome *c* sera, this population is indeed directed against region IV. The second isolated population binds only the guanaco and mouse proteins, which are identical in region VI, and fails to bind the rabbit, cow, and horse proteins, demonstrating that it is directed against region VI. In such simple cases, isolation of the site-specific antibody population is easily accomplished by adsorption on suitable insolubilized antigens (41).

Similarly, mouse cytochrome *c* also differs from the rabbit protein by only two amino acid residues. These differences occur in region II, where the mouse protein has an alanine at position 44 in place of the valine of the rabbit protein, and in region VI, where the mouse, like the guanaco, protein has a glycine at position 89, replacing the aspartic acid of the rabbit cytochrome. Remarkably, mouse cytochrome *c* elicits three antibody populations in the rabbit, one more than would be expected from the amino acid sequence differences. Two of these antibody populations were found to bind to guanaco cytochrome *c*. As can be seen in Table 3, these two populations duplicated the patterns obtained for rabbit antiguanaco cytochrome *c* sera and, on this basis, were shown to be directed against regions IV and VI (42). One explanation for the formation of antiregion IV antibodies against the mouse protein, where there are no differences with the rabbit protein, could be the fact that because region IV is capable of binding antiguanaco region IV antibodies, regardless of whether the region contains a glutamic or an aspartic acid at position 62, it might be capable of also stimulating the corresponding clones. It was found that rabbit antimouse region IV antibodies bind more strongly to proteins that contain glutamic acid in position 62 (guanaco, cow) than to proteins that carry aspartic acid at that residue (mouse, rabbit) (see Fig. 6) (42). Conceivably, the clones capable of recognizing the glutamic acid-containing region IV have not been inactivated, because the binding constants are such that at the very low concentrations at which cytochrome *c* may be normally present in the serum, too few molecules of the aspartic acid containing-cytochrome bind to have any effect. The third antimouse cytochrome *c* antibody population was assigned to region II, because this population would bind only to the mouse protein (Table 3), the only cytochrome *c* used that carries an alanine at position 44.

Because these simple situations are readily analyzed and demonstrate direct correlations between amino acid sequences and spatial structure, on the one hand, and the anti-

Table 3. Stoichiometries of the Reaction of Rabbit Antimouse Cytochrome *c* Antibodies with Various Cytochromes *c*[a]

| Antibody | Cytochrome *c* | | | |
	Guanaco	Mouse	Cow	Rabbit
Antimouse cytochrome *c*	2.09 ± 0.04	2.97 ± 0.11	nd[b]	nd
Antimouse region II	0[c]	0.98	0	0
Antimouse region IV	1.02	0.97	1.03	0.99
Antimouse region VI	1.06	1.04	0	0

[a] The titration of 2.0 ml of rabbit antimouse cytochrome *c* Fab fragments was performed at concentrations of about 10^{-6} M in borate-buffered saline (pH 8.6) with increasing amounts of the cytochrome *c* listed, which ranged from 0.1 to 6.0 nmol.
[b] Not determined (42).
[c] No binding observed.

body populations evoked, on the other, it should be possible to predict on these bases how many antibody populations would be elicited by a particular cytochrome *c* in a particular host and where each population would bind. This task becomes more difficult in the complex situation where the binding of one antibody population interferes with that of another because of the proximity of two or more determinant sites.

Of the cytochromes *c* that are immunologically more complex, only the human protein has been studied enough to allow for a first approximation of the location of its antigenic determinant sites. Human cytochrome *c* differs from the rabbit protein in regions I–VI (Figs. 3a–3d and 4). In region I, the human protein has isoleucine, methionine, and serine at positions 11, 12, and 15, respectively, whereas rabbit cytochrome *c* carries valine, glutamine, and alanine. The region-II difference is at position 44, where the human protein has a proline, while the rabbit carries a valine. The changes from phenylalanine to tyrosine at position 46 and aspartic acid to alanine at residue 50 constitute the differences in region III. The difference in region IV is at position 58, where the human protein has an isoleucine and the rabbit protein carries a threonine. The changes from alanine in the rabbit protein to valine in the human protein at position 83 and from aspartic to glutamic acid at position 89 are the differences in regions V and VI, respectively.

Human cytochrome *c* was found to elicit four simultaneously binding antibody populations in rabbits. Williams and Reichlin (69), who studied the size of the antigen-Fab fragment aggregates by ultracentrifugation, and Noble et al. (68), who determined the stoichiometry of the reaction by fluorescence quenching, showed that *M. mulatta* cytochrome *c*, which is identical to the human protein, except at region IV, binds three of the four antihuman cytochrome *c* populations. It was also found that monkey cytochrome *c* elicits only three populations in the rabbit; it is identical to the rabbit protein in region IV (67). These rabbit antimonkey cytochrome *c* sera were incapable of detecting any difference between the human and monkey proteins. When the rabbit antihuman cytochrome *c* sera were adsorbed with monkey cytochrome *c*, a single antibody population was left unreacted. This antibody population binds only proteins that carry isoleucine at position 58, such as the human and kangaroo proteins (Fig. 4), demonstrating that this population is directed against region IV (67).

The other three determinant sites on human cytochrome *c* have been tentatively located by indirect means. One of the remaining antibody populations could be isolated from the antisera because it is produced in larger amounts than the other rabbit antihuman cytochrome *c* populations. When the antisera are precipitated at equivalence with human cytochrome *c* monomer, a proportion of this population remains in the supernatant fluid (7). This antibody was found to selectively inhibit the reaction of human cytochrome *c* with the cytochrome *c* oxidase of beef mitochondrial particle preparations and was termed the "oxidase site antibody" (70). The site of binding of this population was assigned to region I on the basis of two findings. First, all four antihorse cytochrome *c* antibody populations failed to inhibit the oxidase reaction, as did the antihuman region IV population that binds to the "back" of the protein (Figs. 3a–3d). Horse cytochrome *c* differs from the rabbit protein in regions II–IV and VI. Second, it has been found that modification of lysyl residue 13 in horse cytochrome *c* results in an inhibition of the reaction with cytochrome *c* oxidase (71,72). Although it seems quite probable that this antibody population is, indeed, directed against region I, it should be noted that the specificities of the four rabbit antihorse antibody populations have not been determined and that derivatives in other areas of the molecule can also block the reaction with cytochrome *c* oxidase (73,74). The locations of two more determinants were assigned to regions III and VI (75). However, this assignment was made on the basis of indirect evidence obtained from cross-reactivity studies and not direct binding measurements, and therefore these assignments are tentative. To summarize, the location of one antigenic determinant on human cytochrome *c* has been positively assigned to region IV, there are strong indications that region I contains a second determinant, and the locations of two additional sites have been postulated. More definitive experiments will be needed to positively identify these areas as antigenic sites and to establish whether these are the only determinents on the human cytochrome *c* molecule or whether every one of the six regions at which amino acid sequence variations occur is immunogenic.

This type of approach has both advantages and disadvantages, as compared to other approaches to the location of antigenic sites on the surface of globular proteins. One major advantage is that native intact proteins are used throughout, presenting the appropriate antigenic area in the proper conformation, resulting in binding equivalent to the original immunogen. Initial studies on tryptic and chymotryptic peptides of cytochrome *c* showed no cross reactivity (67). The only exception is one report on the antigenic activity of the heme undeca- and octapeptides of horse cytochrome *c* (76). Such experiments have not been actively pursued with the cytochromes *c*. However, antigenic determinants have been located on several other proteins (tobacco mosaic virus protein, myoglobin, lysozyme, staphylococcal nuclease) by identifying which fragments of the protein, produced by proteolytic enzymes or other methods of cleavage, are capable of cross reacting with antibodies elicited by the intact protein. Many workers have also used synthetic peptide analogs of cross-reactive fragments to study the involvement of individual residues in the observed activity. Both of these approaches have been the subject of many reviews, and extensive explanations of the procedures followed are given (see Refs. 4 and 6).

An advantage of using cross-reactive peptides is that they may give an estimate of the size of the antigenic site (5,6). However, it is not clear if all of the residues of the peptide participate in binding to the antibody or represent a minimum size needed to allow a sufficient proportion of the peptide to assume the proper spatial conformation. It is this type of discrepancy that brings to light the major disadvantages inherent in the use of cross-reactive peptides. First, for globular protein antigens, the peptides employed necessarily correspond to linear segments of the protein, so that one may miss determinants that are formed by two or more remote sections of the amino acid sequence that are juxtaposed in the spatial conformation. Alternatively, the peptides would have to be large enough to assume the conformation needed for binding, in which case they may be a large proportion of the whole protein. Furthermore, large molar excesses of peptide are required to inhibit the reaction, which indicates that the antibody has a much smaller affinity to the peptide as compared to the native protein (6). This difference in affinity may be due to the fact that only a minute proportion of the peptide is in the right conformation for binding and that, under the conditions of these experiments, the reaction with the antibody is insufficient to significantly shift the equilibrium toward the right conformation. Such large molar excesses could conceivably yield misleading results, in that small amounts of impurities in the peptide preparation that may not have been identified may constitute the reactive material.

Many investigators also make use of chemical modifications of proteins to locate the antigenic determinants (77–82). The major disadvantage of this procedure is the difficulty in isolating the modified product in a strictly homogeneous form and in proving that the modification does not affect the conformation of the protein (83–85). If both of these conditions are not meticulously met, the absence of binding to an antibody is uninterpretable. The use of cytochromes c that have a small number of sequence differences is essentially equivalent to using chemical derivatives that retain the native protein conformation and has the advantage that large amounts of reagent are readily available. In contrast, highly purified chemical derivatives are often obtained in very small yields, as compared to the amount of protein used in the reaction (72). In the more complex situations, as with human cytochrome c, the correlations become more indirect, and the identification of determinant sites becomes more tenuous (75). However, as the number of cytochromes c of known amino acid sequence continues to increase, antigenic determinants on more cytochromes c can be identified with certainty, and their antibodies can be isolated.

As discussed in Sec. 3, the amino acid replacements that occur on the surface of cytochrome c do not affect the overall polypeptide backbone conformation of the protein but, rather, produce changes in the surface topography of the molecule. These surface changes are recognized by the immune system, and antibodies of the appropriate specificities are produced. There may therefore be a direct relationship between the number and nature of the amino acid replacements and the immunogenicity of the site. It would be interesting to correlate the properties of these antibody populations to the physicochemical properties of the determinants that elicited them. Only a small number of determinants have been studied in this regard. The findings are presented in the following section.

5. PROPERTIES OF ANTIGENIC
SITES ON CYTOCHROME c

The immunogenic properties of determinant sites on globular proteins can be considered from two points of view. First, how does the structure of a particular determinant relate to the amount and binding affinity of the antibody it elicits? Second, how do the number and nature of antigenic sites on a globular protein affect its overall immunogenicity? Most of the information now available for cytochrome c relates to the first consideration.

Of the six antigenic sites on cytochrome *c* so far studied in any detail, rather than merely localized, only one results from more than one amino acid replacement. This is the region-I determinant on human cytochrome *c*, which contains residues 11, 12, and 15. This determinant elicits the largest amounts of antibody (7). The region-IV site of the human protein is responsible for approximately 25% of the antibody precipitable with the monomeric protein, less than half of the response elicited by region I. On purification, these antibody populations appear to be rather homogeneous, because their fluorescence quenching titration curves are linear with antigen addition. Furthermore, the reaction of these site-specific populations with their corresponding antigens are kinetically quite homogeneous, with "on" constants measured at 0.8×10^6 and 0.45×10^6 M^{-1} sec^{-1} for the antiregion I and antiregion IV populations, respectively (7). All other determinants studied, like the human region-IV site, correspond to single residue replacements. Nevertheless, there is at least a fivefold range in the amount of antibody yielded by such determinants (Table 4).

Of the two antigenic determinant sites on guanaco cytochrome *c*, the site in region IV is clearly the more immunogenic. It is responsible for approximately 60% of the antiguanaco cytochrome *c* response in rabbits (41). Farr titrations showed that this population also binds to the rabbit and mouse proteins but with lower affinity than to guanaco cytochrome *c*. These proteins differ from the guanaco protein in region IV by carrying an aspartic acid at position 62 rather than a glutamic acid (Fig. 6,A) (42). In later bleedings of some rabbits, the specificity of this poplation was found to increase with the appearance of a subpopulation that fails to react with an aspartic acid-containing region IV (Fig. 6,B). The second antigenic site on guanaco cytochrome *c* is located in region VI, where the guanaco protein carries a glycine at position 89, as opposed to the aspartic acid of rabbit cytochrome *c*. The antiguanaco region-VI antibodies were found in smaller amounts and are of slightly lower affinity than antiguanaco region-IV populations (42). Thus, it appears that the introduction of a longer side chain, as in guanaco region IV, results in a greater immunogenicity than does shortening of the side chain, as in guanaco region VI, even though the latter results in an apparently greater chemical difference represented by the loss of a negative charge.

Mouse cytochrome *c*, which is identical to the guanaco protein in region VI, also elicits an antibody population to this site in rabbits. Although the total number of antibody produced against mouse cytochrome *c* in rabbits is only about two thirds of that elicited by the guanaco protein (Table 1), the amount of antibody directed against this

Table 4. Immunogenicity of Single Antigenic Sites on Cytochrome *c*

Antigenic site	Average antibody content of antisera[a] (μg of AbN/ml)	Percentage of total antibody
Mouse region II	7.1	19
Mouse region IV	10.1	25
Mouse region VI	23.2	57
Guanaco region IV	35.4	61
Guanaco region VI	22.6	39

[a] As measured spectrophotometrically (optical absorbance at 287 nm) after removal from specific absorption column.

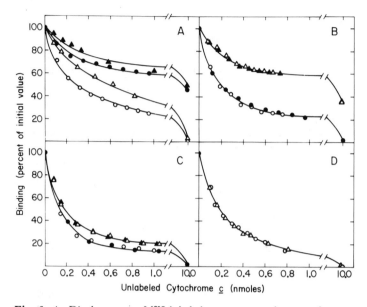

Fig. 6. A: Displacement of ^{125}I-labeled guanaco cytochrome *c* from an early rabbit antiguanaco cytochrome *c* serum, by varying concentrations of unlabeled guanaco (O), mouse (△), cow (●), and rabbit (▲), cytochromes *c* (41,42). Modified Farr procedure according to Nisonoff et al. (67). B: Displacement of ^{125}I-labeled cow cytochrome *c* from a late rabbit antiguanaco cytochrome *c* serum, by varying concentrations of unlabeled guanaco (O), mouse (△), cow (●), and rabbit (▲) cytochromes *c* (42).\C: Displacement of ^{125}I-labeled cow cytochrome *c* from a rabbit antimouse cytochrome *c* serum, by varying concentrations of unlabeled guanaco (O), mouse (△), cow (●), and rabbit (▲) cytochromes *c* (42). D: Displacement of ^{125}I-labeled guanaco cytochrome *c* from purified rabbit antimouse Region-VI antibodies, by varying concentrations of unlabeled guanaco (O) and mouse (△) cytochromes *c* (42).

region is the same for both groups of antisera (Table 4). Furthermore, antiguanaco region-VI and antimouse region-VI antibodies are indistinguishable in their binding to either antigen (Fig. 6,D). Within the narrow range tested, identical immunogenic regions on cytochrome *c* tend to yield the same amounts of antibody that have the same affinity, regardless of the number of other sites on the protein and the response they elicit.

As noted in Sec. 4, region IV of the mouse protein, which is identical to region IV of rabbit cytochrome *c*, is immunogenic in rabbits, and its response may result from the stimulation of clones directed against glutamic acid-containing region IV by cytochromes *c* that contain aspartic acid in the same area. This remarkable phenomenon may also account for the nearly universal cross reactivity of rabbit cytochrome *c* with all of the anticytochrome *c* sera obtained thus far, because all of the proteins tested have such glutamyl-aspartyl exchanges in at least one of their determinants. It may also account for the response observed in rabbits against rabbit cytochrome *c* (34,42).

The third determinant on the mouse protein results from the change from valine to alanine at position 44 in region II. This site was found to be the least effective; it elicited the smallest amount of antibody (Table 4).

6. CONCLUSIONS

The major conclusion that can be drawn at the present level of analysis of antigenic determinants on cytochrome c is that wherever there exists a surface amino acid sequence difference between the immunogen and the protein of the host, antibody populations will be elicited that will specifically react with the area that carries the sequence variation on the immunizing protein and with any other cytochrome c that has the identical structure. For cytochrome c, there is no reason to assume that such external residue changes cause any shift in the polypeptide backbone conformation of the protein, and therefore such determinants are best defined as being purely topographic. Inasmuch as every external residue difference on the cytochromes c studied thus far has resulted in the production of antibody, it should be possible to predict the number and location of topographic determinants on a particular immunogen relative to a particular species, on the basis of the amino acid sequence differences between the antigen and the host proteins and their spatial conformation.

Although generalizations must await the detailed study of more single-determinant sites on cytochrome c, and on other globular proteins, there are several observations that appear to be quite interesting. Identical topographic determinants on the cytochromes c of two different species yield approximately the same amount of antibody, binding with the same affinity, regardless of the immunogenicity of other determinants that may be present on the two proteins (42). Thus, given that the protein antigen is presented in its most immunogenic form, chemically cross-linked polymers in the case of cytochrome c, it appears that antigenic determinants on different portions of the same molecule act as immunogenic sites independently of one another.

These surface amino acid replacements are not equal in their ability to elicit an immune response. Multiple amino acid replacements at a single-determinant site are the most effective (7,70,75), whereas for single residue replacements, there appears to be a definite hierarchy in the responses to various types of replacements. As discussed above for the rabbit, guanaco, and mouse cytochromes c (41,42), the change from an aspartyl residue in the host protein to a glutamyl residue in the immunogen produces a larger antibody response than does the change from aspartic acid to glycine, which, in turn, is more immunogenic than the change from valine to alanine. It remains to be worked out whether, in general, when the difference between the immunizing and host proteins is the introduction of a longer side chain, which produces a protuberance on the surface, a larger immune response results than when the relative difference results from the substitution of a shorter side chain, which produces a depression on the surface of the protein. The effects of such a physical change appear to supersede the possible immunologic effect of chemical differences, such as the loss of a negative charge.

Lastly, the observation that mouse region IV elicits antibody in rabbits, even though it is identical to region IV of the rabbit protein, in that it has aspartic acid at position 62, has been attributed to the ability of this region to stimulate clones directed against it when it contains a glutamyl residue 62. Similar phenomena have been seen in the secondary responses to cross-reacting haptens (86). However, much more work is needed to determine whether this type of phenomenon occurs at other regions of cytochrome c, possibly accounting for the observed rabbit antirabbit protein response (37).

At this time, there seems to be no reason why the approaches used above should not be readily extended to a variety of other cytochromes c and also to other protein antigens. Although these studies have dealt with only the humoral response, future work is not necessarily limited to antibodies. It should also prove possible to study many phenomena related to T-cell activation and genetic control of the immune response against single protein antigenic determinants.

REFERENCES

1. M. Heidelberger and K. Landsteiner, *J. Exp. Med.,* **38,** 561 (1923).
2. K. Landsteiner, *The Specificity of Serologic Reactions,* 2nd edit., Harvard University Press, 1945; reprinted by Dover Publishers, New York, 1962.
3. M. Sela, B. Schechter, I. Schechter, and F. Borek, *Cold Spring Harbor Symp. Quant. Biol.,* **32,** 537 (1967).
4. M. Sela, *Science,* **166,** 1365 (1969).
5. E. Benjamini, R. J. Scibienski, and K. Thompson, *Contemp. Topics Immunochem.,* **1,** 1 (1972).
6. M. Z. Atassi, *Immunochemistry,* **12,** 423 (1975).
7. R. W. Noble, M. Reichlin, and R. D. Schreiber, *Biochemistry,* **11,** 3326 (1972).
8a. E. Margoliash and A. Schejter, *Advan. Protein Chem.,* **21,** 113 (1966).
8b. R. E. Dickerson and R. Timkovich, in P. D. Boyer, Ed., *The Enzymes* Academic Press, New York, 1975, p. 397.
9. M. O. Dayhoff and R. V. Eck, *Atlas of Protein Sequence and Structure,* National Biomedical Research Foundation, Washington, D.C., 1972.
10. D. Borden and E. Margoliash, in G. D. Fasman, ed., *Handbook of Biochemistry and Molecular Biology, Proteins Vol III,* 3rd edit., The Chemical Rubber Co., Cleveland, 1976, p. 268.
11. R. E. Dickerson, T. Takano, D. Eisenberg, O. B. Kallai, L. Samson, A. Cooper, and E. Margoliash, *J. Biol. Chem.,* **246,** 1511 (1971).
12. S. Proger and D. Dekaneas, *Science,* **104,** 389 (1946).
13. D. Keilin and E. F. Hartree, *Proc. Roy. Soc. London,* **122B,** 298 (1937).
14. L. W. Roth, R. K. Richards, and I. M. Shepperd, *Proc. Soc. Exp. Biol. Med.,* **70,** 116 (1949).
15. E. L. Becker and J. J. Munoz, *J. Immunol.,* **63,** 173 (1949).
16. H. Theorell and Å. Åkesson, *J. Amer. Chem. Soc.,* **63,** 1804 (1941).
17. S. Paléus and J. B. Neilands, *Acta Chem. Scand.,* **4,** 1024 (1950).
18. H. Tint and W. Reiss, *J. Biol. Chem.,* **182,** 385 (1950).
19. H. Tint and W. Reiss, *J. Biol. Chem.,* **182,** 397 (1950).
20. E. Margoliash, *Biochem. J.,* **56,** 529 (1954).
21. E. Margoliash, *Biochem. J.,* **56,** 535 (1954).
22. E. A. Kabat and M. M. Mayer, *Experimental Immunochemistry,* 2nd edit., Charles C Thomas, Springfield, Ill., 1971.
23. J. Oudin, *Ann. Inst. Pasteur,* **75,** 30 (1948).
24. J. Oudin, *Ann. Inst. Pasteur,* **75,** 109 (1948).
25. H. Beinert, *Science,* **111,** 469 (1950).
26. J. Storch, R. Tixier, and A. Uzan, *Nature (London),* **201,** 835 (1964).
27. O. Ouchterlony, *Acta Pathol. Microbiol. Scand.,* **32,** 231 (1953).
28. J. Jonsson and S. Paléus, *Int. Arch. Allergy Appl. Immunol.,* **29,** 272 (1966).
29. M. Reichlin, S. Fogel, A. Nisonoff, and E. Margoliash, *J. Biol. Chem.,* **241,** 251 (1966).
30. S. Watanabe, Y. Okada, and M. Kitagana, *J. Biochem.,* **62,** 150 (1967).
31. E. Margoliash, M. Reichlin, and A. Nisonoff, in K. Okunuki, M. D. Kamen, and I. Sekuzo, Eds., *Structure and Function of Cytochromes,* University Park Press, Baltimore, 1967, p. 269.
32. E. Margoliash, M. Reichlin, and A. Nisonoff, in G. N. Ramachandran, Ed., *Conformation of Biopolymers,* Academic Press, New York, 1967, p. 253.
33. E. Margoliash, A. Nisonoff, and M. Reichlin, *J. Biol. Chem.,* **245,** 931 (1970).
34. E. Margoliash, Unpublished observations (1976).
35. Y. Okada, S. Watanabe, and Y. Yamamura, *J. Biochem. (Tokyo),* **55,** 324 (1964).
36. K. Narita, K. Titani, Y. Yaoi, H. Murakami, M. Kimura, and J. Yanecek, *Biochim. Biophys. Acta,* **73,** 670 (1963).
37. M. Reichlin, A. Nisonoff, and E. Margoliash, *J. Biol. Chem.,* **245,** 947 (1970).

38. C. N. Gamble, *Int. Arch. Allergy Appl. Immunol.*, **30,** 446 (1966).
39. D. W. Dresser, *Immunology,* **4,** 14 (1961).
40. D. W. Dresser, *Immunology,* **6,** 378 (1962).
41. G. J. Urbanski and E. Margoliash, *Abstracts, Div. Biol. Chem., 170th Amer. Chem. Soc. Meeting,* Chicago, Ill., 1975, No. 98.
42. G. J. Urbanski, Doctoral Dissertation, Northwestern University, Evanston, Ill, Aug. 1976.
43. M. Reichlin and J. L. Turk, *Nature (London),* **251,** 355 (1974).
43a. M. Reichlin, Personal communication (1976).
44. M. Ultee and E. Margoliash, Unpublished results (1976).
45. E. Margoliash, E. L. Smith, G. Kreil, and H. Tuppy, *Nature (London),* **192,** 1125 (1961).
46. E. Margoliash and E. L. Smith, *J. Biol. Chem.,* **237,** 2151 (1962).
47. E. Margoliash, *J. Biol. Chem.,* **237,** 2161 (1962).
48. W. M. Fitch and E. Margoliash, *Evol. Biol.,* **4,** 67 (1970).
49. E. Margoliash, *Harvey Lect.,* **66,** 177 (1972).
50. W. M. Fitch, *Annu. Rev. Genet.,* **7,** 343 (1973).
51. T. T. Wu, W. M. Fitch, and E. Margoliash, *Annu. Rev. Biochem.,* **43,** 539 (1974).
52. S. B. Needleman and E. Margoliash, *J. Biol. Chem.,* **241,** 853 (1966).
53. W. M. Fitch and E. Markowitz, *Biochem. Genet.,* **4,** 579 (1970).
54. J. M. Fernández-Sousa, J. G. Gavilanes, A. M. Muncio, J. A. Paredes, A. Pérez-Aranda, and R. Rodriguez, *Biochim. Biophys. Acta,* **393,** 358 (1975).
55. A. Lyddiatt and D. Boulter, *FEBS Lett.,* **62,** 85 (1976).
56. G. E. Tarr, Unpublished results (1976).
57. G. C. Hill and G. W. Pettigrew, *Eur. J. Biochem.,* **57,** 265 (1975).
58. F. R. Salemme, S. T. Freer, N. H. Yuong, R. A. Alden, and J. Kraut, *J. Biol. Chem.,* **248,** 3910 (1973).
59. F. R. Salemme, J. Kraut, and M. D. Kamen, *J. Biol. Chem.,* **248,** 7701 (1973).
60. R. Timkovich and R. E. Dickerson, *J. Mol. Biol.,* **79,** 39 (1973).
61. R. Timkovich, E. Margoliash and R. E. Dickerson, *J. Biol. Chem.,* **251,** 2197 (1976).
62. B. Chance, *Ann. N.Y. Acad. Sci.,* **227,** 613 (1974).
63. S. Ferguson-Miller, D. L. Brautigan, and E. Margoliash, *J. Biol. Chem.,* **251,** 1104 (1976).
64. R. Swanson, B. L. Trus, M. Mandel, O. B. Kallai, and R. E. Dickerson, *J. Biol. Chem.,* in press (1976).
65. G. Kreil, *Z. Physiol. Chem.,* **334,** 154 (1963).
66. G. Kreil, *Z. Physiol. Chem.,* **340,** 86 (1965).
67. A. Nisonoff, M. Reichlin, and E. Margoliash, *J. Biol. Chem.,* **245,** 940 (1970).
68. R. W. Noble, M. Reichlin, and Q. H. Gibson, *J. Biol. Chem.,* **244,** 2403 (1969).
69. A. W. Williams and M. Reichlin, *Fed. Proc.,* **27,** 259 (1968).
70. L. Smith, H. C. Davis, M. Reichlin, and E. Margoliash, *J. Biol. Chem.,* **248,** 237 (1973).
71. K. Wada and K. Okunuki, *J. Biochem. (Tokyo),* **66,** 294 (1969).
72. E. Margoliash, S. Ferguson-Miller, J. Tulloss, C. H. Kang, B. A. Feinberg, D. L. Brautigan, and M. Morrison, *Proc. Nat. Acad. Sci. USA,* **70,** 3245 (1973).
73. D. L. Brautigan and S. Ferguson-Miller, *Fed. Proc.,* **35,** 1598 (1976).
74. D. L. Brautigan, S. Ferguson-Miller, and E. Margoliash, Unpublished results (1976).
75. M. Reichlin, *Advan. Immunol.,* **20,** 71 (1975).
76. A. T. Tu and B. S. Hong, *Int. J. Protein Res.,* **II,** 169 (1970).
77. S. Fuchs, P. Cautrecases, D. A. Ontjes, and C. B. Anfinsen, *J. Biol. Chem.,* **244,** 943 (1969).
78. M. Z. Atassi and A. V. Thomas, *Biochemistry,* **8,** 3385 (1969).
79. A. Habeeb and M. Z. Atassi, *Immunochemistry,* **8,** 1047 (1971).
80. M. Z. Atassi and M. T. Perlstein, *Biochemistry,* **11,** 3984 (1972).
81. M. Reichlin, R. Hammershlag, and L. Lavine, in D. M. Weir, Ed., *Handbook of Experimental Immunology,* 2nd edit., Blackwell, Oxford, 1973, Chap. 22.
82. M. Z. Atassi, M. T. Perlstein, and D. J. Staub, *Biochim. Biophys. Acta,* **328,** 278 (1973).
83. B. L. Vallee and J. F. Riordan, *Annu. Rev. Biochem.,* **38,** 733 (1969).
84. A. N. Glazer, *Annu. Rev. Biochem.,* **39,** 100 (1970).
85. G. E. Means and R. E. Feeny, *Chemical Modification of Proteins,* Holden-Day, San Francisco, 1971.
86. H. N. Eisen, J. R. Little, L. A. Steiner, E. S. Simms, and W. Gray, *Israel J. Med. Sci.,* **5,** 338 (1969).

INDEX